T0258384

Measurements for Terrestrial Vegetation

Measurements for Terrestrial Vegetation

Second Edition

Charles D. Bonham

Colorado State University, Fort Collins, Colorado

WILEY-BLACKWELL

A John Wiley & Sons, Ltd., Publication

This edition first published 2013 © 2013 by John Wiley & Sons, Ltd

Wiley-Blackwell is an imprint of John Wiley & Sons, formed by the merger of Wiley's global Scientific, Technical and Medical business with Blackwell Publishing.

Registered office: John Wiley & Sons, Ltd, The Atrium, Southern Gate, Chichester, West Sussex, PO19 8SQ, UK

Editorial offices: 9600 Garsington Road, Oxford, OX4 2DQ, UK
The Atrium, Southern Gate, Chichester, West Sussex, PO19 8SQ, UK
111 River Street, Hoboken, NJ 07030-5774, USA

For details of our global editorial offices, for customer services and for information about how to apply for permission to reuse the copyright material in this book please see our website at www.wiley.com/wiley-blackwell.

Library of Congress Cataloging-in-Publication Data has been applied for

ISBN 978-0-4709-7258-8 (hardback)

A catalogue record for this book is available from the British Library.

Wiley also publishes its books in a variety of electronic formats. Some content that appears in print may not be available in electronic books.

Cover image: From Shutterstock, by Markus Gann: An image of the australian rain forest
Cover design by Steve Flemming

Set in 10.5/13pt Times by Aptara Inc., New Delhi, India

1 2013

This book is dedicated to

Aaron Paul Burdick and Gregory Scott Burdick.

My identical twin grandsons who gave me immeasurable love and inspiration.

Contents

Preface

The second edition of this book reflects a change in the approach taken in development of the first edition. In that edition, I attempted to simply present information on vegetation measurements as reported in published papers, reports, and existing books at that time. Included was any study that used one or more of the four characteristics of plants as they occur in communities, associations, in stands or other units of vegetation categories. I made very few comments on studies that compared data results obtained from various equipment to measure cover for example. I found very few published papers or reports of such comparisons that provided sufficient details that the study could be repeated by another researcher. In particular, study conclusions did not report statistical evidence as to which method or equipment was found to be the most precise, reliable, and/or repeatable, even in the same study area.

It is recognized that reported studies of comparative estimates of a measure are very old. Yet, it is a fact that budget managers in natural resource offices have never given priority to adequately finance such studies; this is so, even as needs for use of vegetation measurement data in decision-making changed over the decades. It follows that the existence of new methods for measurements are few in published literature. This new edition still contains a few old evaluations of method comparisons of plant cover estimates obtained from different quadrat sizes and shapes. These are included to suggest size and shape of plots that modern day field ecologists might consider for preliminary evaluations before embarking on a full-scale vegetation study. In contrast, old habits and traditions die hard, allowing professional plant ecologists in academics, private industries, and state and federal agencies to continue vegetation measurements of attributes that are not based on sound scientific principles. The use of prior methods and procedures to collect and analyze field data continues with indifference as to whether "old methods" result in biased, non-repeatable, and imprecise estimates. Almost any of these latter problems can be mitigated by proper sampling designs and subsequent data analysis.

The greatest problem in obtaining a measure often is not the equipment used, but rather the data analysis used to generate data summaries. Statistical methods for data analysis have increased at a relatively rapid rate, while measurement equipment has essentially remained the same. Yet, this book is about the measures themselves, not how to analyze the data collected. Even so, some newer methods of data analysis, especially those involving spatial statistics, have been included in this new edition, but only in brief form. Data collection methods were added to include spatial statistical sampling designs and examples for analysis are provided from both published and unpublished sources.

No attempt has been made to give more concise meanings of words and terms than those already employed for up to a half-century. These include the use of double sampling in place of regression equations, quadrat for plot, and line transects from a measurement tape, to mention a few. The book deals with principles and procedures used to obtain structural measurements of terrestrial vegetation communities. Emphasis is placed on frequency, cover, density, and biomass, as commonly defined in the preponderance of literature on such measures.

I am grateful to the reviewers of the manuscript. Their editorial as well as other suggestions offered improvement to the manuscript. I am especially indebted to Ms. Margaret Broadbent who provided the expertise needed to develop figures, tables, and special formats for the text. Her attention to detail needed for each revision and subsequent finalization of each chapter is greatly appreciated. I extend my sincere appreciation to the editorial staff, especially to Rachel Wade, Izzy Canning, and Fiona Seymour, for their forbearance and kindness extended me throughout the entire process at John Wiley & Sons, Ltd, United Kingdom.

Charles D. Bonham
Colorado State University
Fort Collins, Colorado

About the companion website

This book is accompanied by a companion website:

www.wiley.com/go/bonham/measurements

The website includes:

- Powerpoints of all figures from the book for downloading
- PDFs of tables from the book

1

Introduction

Variation in the morphology of plants resulted in the grouping of plants into broad categories on the basis of life-forms. Major life-forms are represented by terms such as "tree," "shrub," grass," and "forb." These life-forms often provide a basis to describe major terrestrial plant communities (Odum and Barrett 2005). Life-forms of plants or plant species can be described by a number of characteristics such as biomass, frequency, cover, and density. Some life-forms or plant species are, perhaps, better described by certain characteristics of measure than are others. A combination of objectives of a study and species involved will determine what characteristics are to be measured for an effective description of vegetation. To see this more clearly, consider the measurements of cover, which are estimates of relative areas that a plant controls to receive sunlight. In comparison, biomass directly indicates how much vegetation is present, and particular species indicate the amount of forage available to herbivores in the area. Density describes how many individual stems or plants occur per unit ground area, while frequency describes the dispersal or distribution of a species over the landscape. Each of these species characteristics has a distinct use in vegetation characterization and description. Often, several measures are used in combination for an in-depth description of vegetation (Bonham 1983).

Frequency, cover, density, and biomass are expressed in quantities per unit of area and these are units associated with equipment. Points, plots, and tape measures are often used to obtain these quantities of measure. On the other hand, a measurement technique includes the process used to obtain the measure, specifically location of the observation, clipping, observing a hit by a point, and summarization of the data. Thus, methods used in measurement are ways of doing things to obtain the measure. Often the terms "methods" and "techniques" are used synonymously in the vegetation literature to include pieces of equipment. Thus, one reads phrases such as "the method used was a 0.5 m^2 quadrat." No distinction is made here between techniques and methods because both describe or imply a process used in obtaining a measure of given vegetation characteristics.

Measurements of vegetation characteristics have been made for more than a century, and methods developed to obtain these measurements are numerous. Few

Measurements for Terrestrial Vegetation, Second Edition. Charles D. Bonham.

methods are comparable, even for measuring the same characteristic of vegetation, such as cover. This is true because objectives to obtain the measure differ. Yet, comparisons of vegetation characteristics over time and space are often necessary. If comparisons are to be made, then comparable sampling methods must be used to obtain the measure. For instance, if 100 plots of 0.1 m^2 area are measured for cover, then 10 m^2 of area have been measured. To compare these results to the use of a 50 m length fine point transect, the number of transects measured has to be associated with an area of 10 m^2. The same area measured would then be comparable (Bonham and Clark 2005). More on this is presented in the individual chapters on measures.

There are basically two objectives to be considered in the selection of measurements and methods used to obtain measurements. For example, one objective may be to describe the characteristics of the native vegetation at a given time period, and then again at a later date, to assess changes. The second objective is to describe vegetation characteristics so that the measurement can be used as a standard or a baseline. Sampling intensity has to be the same for both time periods; that is the sampling process has to sample the same area although the equipment area has changed. Otherwise, all descriptions of the vegetation would be relative to different areas measured and not subject to valid comparisons. The two objectives are compatible and should not be considered as competing for resources, especially monetary, in order to be attained.

The purpose in a study of methods to obtain vegetation measurements is twofold in nature: (1) to make proper choices of appropriate methods used to estimate a characteristic of vegetation, that is, frequency, cover, density, and biomass, and (2) to properly select and utilize a sample design that will provide unbiased estimates of the characteristic measured. One will gain efficiency in carrying out both of these objectives simultaneously only through proper use of methods in the field, followed by effective data analysis procedures.

No known set of techniques is free of disadvantages for any measurement or set of measurements. Rather, selection of a technique should be made with perhaps an understanding that certain modifications may be needed to optimize its use. Modifications of equipment to reduce effects of biased estimates or to limit disadvantages are frequently described in the literature. In general, however, all methods used to measure vegetation are related to land areas. That is, frequency, cover, density, and biomass amounts are correlated not only for certain species but also by the fact that each is estimated with reference to a land area and, in some cases, volume occupied.

1.1 Historical brief

Measurements of vegetation date to antiquity. In the third century B.C., Theophrastus observed that certain relationships existed between plants and their environment.

Thus, he was an early contributor to plant ecology in the quantitative sense. Still, centuries passed with qualitative assessments dominating vegetation descriptions, and plant species in particular. Geographic descriptions of vegetation occupied the interest of many naturalists in Europe, where only listings of species dominated their efforts. Indeed, lists of plants provided the beginning of vegetation characterization as a true quantitative approach. Emphasis was placed on such lists throughout the eighteenth and nineteenth centuries on the European continent. It was in Europe that Raunkiaer (1912) used the first known plot (0.1 m^2) to obtain a quantitative measure of plant species frequency, although the work was still to describe geography of plants (Raunkiaer 1934).

The work of F. E. Clements in the United States increased precision in vegetation measurements. Clements, in 1905, coined the word "quadrat" for use in vegetation data collection. While the term technically defines a four-sided plot, its usage over time has been adulterated to include any plot shape, even a circle.

Gleason (1920) further advanced the concept of quantitative measurements. He described applications of the quadrat method in description of vegetation characteristics. In particular, Gleason (1922, 1925) developed a thorough explanation of species and area (or space) relationships. His concepts led to the idea that sample adequacy could be determined from the number of species encountered as a plot area increases. The well-known species–area curve is still used at present to determine relative sample adequacy. There are limitations to this method of determination of proper plot size; namely, each species would have a different size of plot.

Measurements and analysis of vegetation characteristics during the 1920s led to statistical applications to plant ecology. Kylin (1926) introduced the concept of "mean area" and defined it to be the inverse of density, that is, the number of individuals per unit of area. He was also among the first to present explanations for the relationship between density and frequency, which is of a logarithmic nature, not linear. Furthermore, species absence, not presence, determines density. Kylin's work followed that of Svedberg (1922) in Europe, and their approach to measures of vegetation was of a statistical nature that encouraged many others to examine vegetation characteristics from a quantitative point of view.

Cain (1934) and Hanson (1934) compared quadrat sizes, while Ashby (1935) gave an early introduction for the use of quantitative methods in vegetation descriptions and Ashby (1936) published on the topic of statistical ecology. Bartlett (1936) gave examples of statistical methods for use in agriculture and applied biology, but Blackman (1935) had previously introduced statistical methodology to describe the distribution of grassland species. These early studies in vegetation measurements emphasized the dispersion of individuals in plant communities. Thus, patterns of dispersal were very much in the forefront of most quantitative assessments of vegetation characteristics.

How plants are arranged spatially implies distance measures and, subsequently, pattern. This emphasis on pattern analysis began in earnest in the 1920s and reached a peak in the late 1940s and early 1950s (Greig-Smith 1983). Interest in species patterns was briefly rekindled in the 1960s as distance measures were again used for density estimates (Green 1966, Beasom and Houcke 1975). Many distance measures in plant patterns use a single, linear dimension, and distance measures have been referred to as "plotless" methods. The return to plotless methods in the United States was essentially driven by time–cost considerations needed for large-scale inventories of forests and rangeland resources. Recently, these methods have been referred to as "variable area" methods and this description for these methods is used here because, indeed, "area" is involved in all of them.

Since distance measurements included rigid assumptions about the distribution of individual plants, an understanding of patterns found in natural plant populations was necessary. Therefore, a great deal of effort was expended to develop acceptable modifications to variable area methods for use in the estimation of frequency, cover, density, and biomass. Thus, measurements of vegetation actually began to find a place in the work of professional plant ecologists from the 1920s onward. Today, many professionals in vegetation ecology have mastered the seemingly more difficult merger of this discipline with that of statistics.

1.2 Units of measure

The science of measurement, which is called metrology, has been a vital part of science, especially the physical sciences, for centuries. The science of metrology was given much attention during the nineteenth century because a better system of units and standards for measurements was needed to assist the field of physics (Pipkin and Ritter 1983).

The metrology of vegetation itself, however, is of even more recent origin. In the decade 1970–1979 there was major progress on the determination of fundamental constants needed to relate measured vegetation characteristics displayed by density, cover, and so forth to biological and ecological theory.

In the decade of the 1960s, the International Biological Program (IBP) introduced the integrated systems approach to study the interrelationships of organisms and their environment that operated in an ecosystem. Mathematical and statistical models formulated up to the present time have provided fundamental insight as to how such systems of organisms functioned individually and collectively. Thus, for example, constants for energy and nutrient transfer through a system were provided, which resulted in a clearer understanding of how measured characteristics described plant–environment relationships. For example, the amount of biomass accumulation by an individual species can be used to assess that species' role in nutrient utilization and recycling within the vegetation system as a whole.

Most vegetation measurements are now made in metric notation, which is used throughout this book. Table 1.1 provides a definition of the relationship that exists

Table 1.1 Metric weights and measures

Linear measure	
10 millimeters (mm)	= 1 centimeter (cm)
10 centimeters	= 1 decimeter (dm)
	= 100 millimeters
10 decimeters	= 1 meter (m)
	= 1000 millimeters
10 meters	= 1dekameter (dam)
10 dekameters	= 1 hectometer (hm)
	= 100 meters
10 hectometers	= 1 kilometer (km)
	= 1000 meters
Area measure	
100 square millimeters (mm^2)	= 1 square centimeter (cm^2)
10 000 square centimeters	= 1 square meter (m^2)
100 square meters	= 1 are (a)
100 ares	= 1 hectare (ha)
	= 10 000 (m^2)
100 hectares	= 1 square kilometer (km^2)
	= 1 000 000 (m^2)
Volume measure	
1000 cubic millimeters (mm^3)	= 1 cubic centimeter (cm)
1000 cubic centimeters	= 1 cubic decimeter (dm^3)
	= 1 000 000 (mm^3)
1000 cubic decimeters	= 1 cubic meter (m^3)
1 000 000 cubic centimeters	= 1 000 000 (mm^3)
Weight	
10 milligrams (mg)	= 1 centigram (cg)
10 decigrams	= 1 gram (g)
	= 1000 (mg)
10 dekagrams	= hectogram (hg)
	= 100 (g)
10 hectograms	= 1 kilogram (kg)
	= 1000 (g)
1000 kilograms	= 1 metric ton (t)

among linear, area, volume, and weight measures from the metric system. The volume measure is given in this form because some measures of weight of vegetation biomass should be reported as per unit volume occupied. Essential or fundamental constants used in measurements of vegetation may include conversion from one system to another. For example, in order to interchange units from the metric to the English system, constants are needed. Approximation values for some of these constants are given in Appendix Table A.1 (Appendix at end of the book).

Additional constants to those in Table 1.1 are often needed in vegetation measurement work. Such constants are given in Appendix Table A.2 since these small areas are often used to estimate weight of plant biomass, especially that of forage, in much of the world. Vegetation workers on rangelands of the United States and other places often use pounds of forage per acre (lb/acre) to manage forage resources, yet use metric dimensions for obtaining these estimates. Therefore, conversions are given for the most commonly used plot areas.

Calculation of a constant for any given area to a larger area is as follows:

$$\text{unit of wt(1)/large area} = \frac{\text{one unit of wt(2)/large area}}{\text{(unit of wt(2)/unit wt(1)(small area))}} \tag{1.1}$$

That is, for a 0.25 m^2 plot (Appendix Table A.2),

$$\text{kg/ha} = \frac{\text{g}/10\,000\,\text{m}^2}{(1000\,\text{g/kg})\left(0.25\,\text{m}^2\right)} = \frac{10}{0.25} = 40$$

Still other constants may be useful for conversion of one or more measures into another measure. For example, cover percentage of a species may be used to estimate biomass weight (grams) of the species, in which case the constant is estimated by least squares procedures of regression analysis. The general form of the equation is usually

$$\text{Biomass (g/area)} = f\left[\sum_{i=1}^{k} (\text{measure } i)\right] \tag{1.2}$$

where i equals one or more independent measures such as cover, stem diameter, and so on, and k is the number of independent measures made on the plant. The function f also includes measurements in the equation as given in appropriate chapters of the book.

1.3 Choice of method

Selection of the proper method to obtain a measure is based on several criteria. Emphasis is based on two major characteristics: (1) those involving the physical aspects of vegetation, and (2) those involving biometrics and econometrics of the methods.

1.3.1 Vegetation characteristics

Selection of proper measurement methods to use requires knowledge of the floristic composition of the vegetation type. That is, equipment to estimate density, biomass, and cover is determined by the plant community life-forms. The abundance of these measures determines, for example, the size of quadrat, the specific distance measure, and the number of sampling points needed. In general, dense vegetation, which usually implies higher density of individual plants, larger plants, or both, can be measured with fewer large plots or fewer points than a sparsely occupied area. The latter areas often require more points, plots, and so on because the variation is larger for the measure, such as cover or biomass.

Life-forms present in a vegetation type are suggested as a consideration in selection of a method. But sometimes a given form, such as the shrub or tree form, will have more variation in size over species than will grasses or forbs in a vegetation type. The same can be true for forbs compared to grasses. That is, size variation among species within a life-form can suggest the use of a method such as the size of a quadrat for individual plant counts or biomass determination.

Spatial exclusion, by large plants, or other species from a given size of plot area, might be overcome by increasing the sample size, in which case, sampling design should be considered in an application of adaptive sampling (Chapter 4). Patterns of distribution for individuals within a species, and patterns for species, are of major concern in method selection and use. If all distributions were random, or very nearly so, then measurements would not yield biased results in most cases. Random distributions imply that no pattern is present in the measure obtained and that such measures can be obtained from a random sampling process to provide unbiased estimates of the measure. In vegetation measurements, the presence of patterns in the measure causes the greatest biometric concern, which, in turn, influences the economics of measurements.

1.3.2 Biometrics and econometrics

Biometrics (the science of statistics applied to biological observations) and econometrics (the science of statistics applied to economic data) are useful in the selection of appropriate measurement methods. Statistics provide estimates of the population mean and an estimate of its variance with assignable probabilities for confidence limits. When the data distribution form (e.g., normal or binomial) is known or assumed for a measure, a method that provides the smallest value of an estimate of the variance of the mean is the best estimate of the mean. Furthermore, such a method provides the minimum number of observations to be taken for that given measure of biomass, cover, and density.

Obviously, the foregoing discussion implies that certain methods will also be more efficient from the economic viewpoint. Any method that requires more observations than another to obtain the same precision is not as cost-effective if the cost is the same for each observation. Additionally, precision of the estimates is to be considered. In some cases, a method may be precise, that is, gives repeatable sampling results. However, it may not be accurate, which means that the method does not estimate the true population value very well.

Emphasis should be placed on the sampling error of a method. This error is not a mistake or an oversight, but rather involves the variation present in actual measures (cover, density, and biomass). An estimate of this error is usually defined by the "standard error" of the mean. The magnitude of the sampling error depends on: (1) the number of observations, (2) the inherent variability of the measure, and (3) the method of selecting a sample. That is, location of sampling units in the field is made according to random or other methods. All of these aspects of the sampling error will affect the cost of sampling.

1.4 Variation in vegetation

Sources of variation in measures of vegetation characteristics are many. Vegetation characteristics of frequency, cover, density, and biomass are affected by species life-form, species composition, seasonality, previous use by humans and animals, and edaphic (soil) and climatic characteristics. Already, life-forms have been suggested as a variable that affects the selection of a method used for a measure. Life-form (tree, shrub, and herb) is a source of variation in vegetation in terms of its relationship to frequency, amounts of cover, number of individual plants possible in an area, and effects on biomass of a plant.

Precipitation, air and soil temperature, soil moisture, and time in relation to initiation and cessation of plant growth all affect the measure of a characteristic for a certain species. More importantly, however, they contribute significantly to variation of the measure when made over all species present. For example, total biomass of an area depends on the phenological stages of the species present during sampling. Variations among time intervals over a growing season or years, then, will depend on the development stage of both major and minor species. Previous use of, or destruction of, vegetation is often reflected in the variation found in vegetation characteristics. Prior harvesting of trees, heavy grazing by large herbivores, periodic infestation of insects, and disease are significant sources of variation in measures of vegetation. Variations are often noted also by species composition, which indicates a secondary or greater successional stage of the plant community. Edaphic sources, as contributors to variation in vegetation, include parent material, stage of soil development, and physical and chemical properties of the present soil.

While weather is a measure of present events such as air temperature, wind, and precipitation, climate is a long-term phenomenon. Weather, in general, affects the measure of vegetation cover more so than it does density for perennial species. This is true for established individual plants of a species. In the case that one is interested in variation in life cycles of one or more species, density counts should be made on an annual basis to document life stages of the species from germination to maturation over the growth season. However, annual and ephemeral species are affected through expressions of frequency, density, cover, and biomass by weather events occurring within a few days. Climate, on the other hand, introduces variation

Table 1.2 Relations of vegetation characteristic variability and measurements of cover, density, and biomass

From small	Variability	To large
	Sample size increases ⇨	
	Plot size increases ⇨	
	Need for stratification increases ⇨	
	Effectiveness of double sampling increases ⇨	

in vegetation through its determination of species composition and reproduction in perennial species. These sources of variation should be used to stratify for sampling purposes to enhance efficiency in the sampling process. A stratum should be used in the sampling design to classify measurements to ensure minimum variability within a stratum and maximum variability among strata for measures of interest. Vegetation typing into plant communities provides a more efficient estimate, both statistical and economical, of the measure. Season of growth for herbaceous plants should be stratified into early, mid, and late season, while kinds of past use and intensity of use also provide strata. Table 1.2 illustrates some general effects that variation in vegetation characteristics have on sampling the measures.

1.5 Observational units

A unit is a distinct, discrete member of a population that can be analyzed in an aggregate or as a whole. It is any quantity (weight, percentage, length) used as a measurement and is representative of the population as a whole. Then, amount of plant biomass in an area of a plot is a single unit, just as a point touching a leaf is a unit of cover. Thus, some literature references prefer to use the term "sampling units" that is used in this book.

1.6 Sampling

Sampling is best understood when compared to a complete enumeration or census. Both of the latter refer to measurements of every individual unit in a population. For example, every possible quadrat or distance between plants constitutes a population. It is well known that a sample from a population of measurable units will essentially contain the same information as a complete enumeration, yet will cost far less than the census. What is less well known, though, is the fact that sample-based data may be more reliable than a 100% inventory. This follows from the fact that samples are often taken with greater care than can be used in a complete census because more expertise can be used in sampling. Sample-based data also can be collected and processed in a small fraction of the time required for a complete census. Even in a census, there are, in fact, non-sampling errors occurring. These include missing observational units, double-counting, etc. that can make the census less accurate than a sample. The objective for sampling is to obtain an unbiased estimate of the population parameters, namely, the mean and its variance. Samples are made up of a number of observations. Each sample is of size n, the number of observations made for the measure of interest. Then, each sample has a set of observations specific to it. Therefore, the word "sample," in the statistical sense, refers to a set of observations, not a single observation. An observation, in contrast, is a single measurement made from a sampling "unit" defined in Section 1.5 and Chapter 4. This definition should be used to prevent misinterpretation of sample size information.

An understanding of basic statistical concepts should be developed before fieldwork is begun, regardless of the reason for obtaining measurements. For this purpose, principles of statistics are given in elementary detail in Chapter 3. Of particular importance is the relationship of variation and sample size needed to adequately describe population measures.

Concepts of accuracy and precision are important in the selection of a measurement unit. For example, some sampling units may be precise (results are repeatable) but not necessarily accurate (providing true population value). The difference between the two emphasizes the sample estimate (precision) while "accuracy" refers to the true population measure that cannot be known for a sample.

Sampling units used to measure characteristics of vegetation influence the data distribution. For example, the size of a plot determines the number of individuals that can occur within the plot boundaries. Then, the Poisson distribution can be derived from consideration of plot area relative to the total area sampled. No derivations are given in Chapter 3, only explanations of the relationships of measurements to distributions and relationships among distributions. The Bernoulli distribution is discussed to develop an intuitive basis for occurrence of four binomial distributions. These distributions can be useful in studies of individual species such as threatened and endangered species, invasive species, and indicator species. The point-and-count binomial distributions are given with parameters and associated sample size adequacy as estimated from an equation based on the normal distribution.

1.7 Frequency

Frequency is the percentage of a species present in a sampling unit and, therefore, it is influenced by size and shape of sampling units. Frequency can be estimated by plots only because the measure is based on occurrence per unit area and by definition, the point eliminates area dimensions (Chapter 5). Frequency is a useful index for monitoring changes in vegetation over time and comparing different plant communities. Mathematical relationships may be used to estimate plant density and cover from frequency data. The index is also highly sensitive to abundance and pattern of plant growth. Selection of an appropriate plot size and shape requires preliminary study of the vegetation type. Some plant ecologists recommend plots of certain size on the basis of experience, while others calculate plot sizes from mathematical relationships between some vegetation characteristics for statistical considerations. The scope of frequency and details for calculation of plot size, plus other considerations are discussed in Chapter 5.

1.8 Cover

Cover can be measured for vegetation in contact with the ground (basal area) or by projected aerial parts of vegetation onto the ground (foliage cover). All methods of measuring cover depend on occurrence of a plant part within a quadrat or by contact with a pin, the area of which may be very close to zero (0). In fact, all methods give cover estimates on a two- or three-dimensional basis. Yet, a line is often erroneously assumed to be one-dimensional when, in fact, its width is very narrow, while its length is very long relative to this width (Bonham and Clark 2005). It is important, as stated above, to realize that all measures are in units of area sampled; cover estimates from point or line intercepts are no exceptions.

Points have been used singly, as in the step-point method, or in frames, as in the point-frame method. Spacing of pins within a frame is, in general, closer for intensive studies, and farther apart for general surveys. This is so because individual species are of more interest in the former, while a general cover estimate will suffice in the latter. Cross-hairs in telescopic sights are used for sighting the points rather than the lowering of a pin. The line-point method involves point readings spaced along a line-transect; spacing of points depends on the size of the study area and density of ground cover. In no case should there be more than 100 points for a sampling unit. The difficulty with a biological-ecological interpretation of species cover values less than 1.0% is obvious. Furthermore, individual line-point transects to obtain estimates of basal cover of plant species have been shown to be unreliable and produce biased results because individual transects cannot be precisely relocated in the exact position with no deflection angle from that point (Bonham and Reich 2009).

Line-intercept often uses a tape measure as a line-transect through a plant community of interest and measuring the length of a species basal or foliage area intersected. Line-intercept methods give values closer to true cover as calculated from ellipse formulas than with variable area plot methods. Line-intercept data are more precise, and data are obtained more rapidly than by the use of quadrats in communities with different-sized individuals of plant species. It is a good method only for a few types of situations, e.g., measuring the canopies of shrubs or mat-forming plants, or measuring the area covered by different plant communities. Relocation of line-intercept transects for monitoring plant species changes suffer from the same errors as the line-point transect does and should follow recommendations provided for latter methods to avoid bias or at least minimize it.

Quadrats of varying sizes have also been used to measure cover. The most often used quadrat methods involve ocular estimates of percentage cover by species. Cover classes and use of the mid-point value of the cover class for data analysis usually accomplish this. Use of cover classes enables repeatable estimates to be made by different observers and the data cannot be analyzed by standard statistical methods. Some simulation studies on mid-points of cover classes have shown that there is small bias in means of these mid-points over a study area compared to means estimated by standard statistical methods. However, non-sampling errors by observers can be large for cover class methods or by methods to obtain estimates to the nearest 1.0%.

Variable area plot techniques have been used to estimate shrub cover, but not extensively – specifically, because most methods are more difficult to use as shrub cover increases. Variable area plot techniques include distance measures and other unbounded units. On the other hand, variable plot techniques might be more efficient than other methods (bounded as a quadrat) in open shrub and forested communities. Yet there is another method to obtain unbiased estimates of shrub cover given in Chapter 6 and is referred to as the line shadow transect (LST) (Thompson 1992). In essence it is an intercept transect, but differs in the method of obtaining estimates of the canopy cover. The method uses the placement of several line transects (sampling units) within a plant community of dominant shrub species. Measurements of the major and minor axes of an ellipse-shaped area of canopy are made on intercepted shrubs and these canopy areas are projected onto a baseline transect from which sample transects originate. Equations to obtain unbiased estimates of the mean and variance of areas covered by shrubs within the plant community are provided.

1.9 Density

Density measurements require that a defined individual unit of a plant be countable. Density, historically, has been an important measurement for trees and shrubs, and somewhat less important for bunchgrasses and forbs. Essentially, the counting process is slow and tedious for herbs and is questionable for use with sod-forming grasses and other growth-forms where the identity of an individual plant may be difficult to establish. However, density can be estimated if a standard counting unit is defined properly. Plots of various bounded areas and variable area methods to estimate density include: plots of various shapes and areas, point-centered quarter

(PCQ), wandering quarter, line transect, random pairs, nearest neighbor, and closest individual. These latter variable area plots are considered to be minimally biased, if not unbiased. Other often used methods to obtain estimates of density provide biased estimates.

1.10 Biomass

Biomass is another primary vegetation measure because it indicates the quantity of resources, such as water, used by plant species in the community. Biomass is a relative measure of a community's resources bound up in different species. The primary method used to measure herbaceous biomass is to clip biomass in quadrats. Tree and/or shrub biomass is usually estimated from dimensional analysis. That is, other measures, such as plant height and crown diameter, are made and used to predict biomass from regression equations.

Historically, long narrow plots, placed across the direction of change, were generally considered the most efficient, but recent studies contradict this view. It can be shown that proper sample design and number of sampling units provides unbiased estimates of biomass (or other measure), irrespective of spatial patterns.

Optimum plot sizes may differ for determination of biomass of plant species. Additionally, optimum size and shape of plot will differ among vegetation types. Furthermore, sample adequacy, in terms of observations needed, also varies according to the size and shape of the quadrat, species combination, and vegetation type. To appreciate why this is true, recall the sources of variation in vegetation characteristics presented in Section 1.4.

Biomass of shrubs is usually measured with two major components in mind: total biomass and forage for animals. If the objective of a project is an ecological characterization of the vegetation, then only total biomass need be considered. Otherwise, if large herbivores are present and feed on shrubs, then forage (or "browse" as it is often termed) is measured. Shrub and tree biomass and/or current annual growth are determined by indirect methods using dimensional analysis. As previously suggested, the most frequently used dimensions include crown diameter, crown area, height, and basal stem diameter.

Shrub foliage estimates and browse are based on twig diameter, length, and count measurements. Then, regression equations are computed by size classes for shrubs and vegetation types. In fact, very few studies have shown meaningful relationships between tree and shrub size and biomass unless plant size classes by site are considered. Site differences probably account for growth differences and express environmental differences directly. Suggested measurements for double sampling of shrubs and trees are crown length and width, plant height, basal stem diameter, and percentage of live crown cover. The percentage cover may not be useful for indirect estimation of total biomass, but in species with large patches of dead crown, cover provides an acceptable estimate of biomass when used in regression equations. Chapter 8, on biomass, provides more details.

1.11 Measurements with remote sensing

Earth-imagery satellites have not yet provided methods to obtain the precise vegetation species measurements needed in most plant ecological projects. However, spectral imagery from aerial and satellite platforms can be used effectively to describe vegetation-environmental systems as a combination within pixels produced by the spectral imagery. Such systems can be related to ground surface areas via pixels generated with their coordinates. These pixels can be ground located within a reasonable precision level, sampled by use of quadrats, lines, and/or points to obtain a description of species composition, biomass, density, and/or ground cover. The plant ecological emphasis is on the pixel area and the vegetation-environment of that 900 m^2 area (less than 0.25 acre). Relocation of an individual pixel with a given precision level depends on the quality of the GPS used. Often there is a group of pixels having the same combination of spectral bands and, in such a case, plant communities can be mapped within the group.

There is considerable literature on forested vegetation types where satellite imagery has been used to map tree crown cover in general and in relation to insect and fire damage incurred over large areas of forests. Aerial and satellite remote sensing have been used to map rangeland types and to detect and monitor some invasive plant species. But information on plant species cover, density and biomass has continued to be collected by traditional "on-ground" methods.

There is some discussion on sampling designs to locate similar combinations of spectral imageries in pixel size (30×30 m) areas. Field designs to collect detailed information on vegetation-environmental combinations within these pixels are easy to implement and relocate for monitoring changes over time. Emphasis might be on a total assessment of species combinations and their environmental factors (topography, soil, climate, past use history, etc.) present in association with individual pixels of a scene or partial scene.

Methods of measures using remote sensing and spatial sampling designs are presented in the chapters on cover, density, and biomass. Uses of these general methods are also presented in the chapter on monitoring and evaluation.

1.12 Monitoring and evaluation

Objectives of vegetation measurements differ with respect to various project goals. One may be interested in mapping plant communities and studying changing patterns in dominant species as they interact with changes in soil–water relationships. The objectives and methods of vegetation measurement may be different, but the plant characteristics to be measured remain the same: frequency, cover, density, and biomass. Once information is available on these characteristics, data can be synthesized and analyzed in a manner that best suits the objectives of the study.

A prerequisite of any vegetation measurement for inventory purposes is general familiarity with the area and recognition of general physiognomy of the vegetation. This can be achieved through a reconnaissance survey of the area. Topographic maps, soil survey maps, and aerial photographs are very helpful in such a survey. These maps can be overlaid to study topographic and soil influences on plant characteristics and species composition. It is, therefore, possible to map plant communities and their descriptions on survey maps. Once these communities or ecological units are delineated, the next step is to choose a sampling design and a measure of plant characteristics. A summary of the sampling statistics should be used in a narrative description of the ecological sites. These statistics should include spatial terms to account for autocorrelation of individual species of importance.

There are few areas not grazed by domestic or wild animals. In fact, competition between domestic and wild animals is significant in many parts of the world. Moreover, herbivorous insect populations sometimes reach epidemic proportions that have detrimental effects on native vegetation. Even at normal population levels, small herbivores such as insects, although sometimes not visible, consume substantial quantities of forage that would otherwise be available for large herbivores (Capinera 1987).

Methods of determining carrying capacity for herbivores range from general reconnaissance to detailed surveys. The choice depends on objectives and resources available to undertake such a survey. Simple information on herbage biomass is not a sufficient basis for decision-making in regards to number and kinds of herbivores that might graze on a given ecological site or plant community type. Such a decision depends on the condition (health) of the vegetation relative to a natural environment. The objective is usually to maintain vegetation resources in at least a sustainable condition for grazing of large herbivores. Therefore, monitoring to detect any change of plant species composition enables adjustment of land use practices when necessary. Planning and management decisions for vegetation use by humans and animals should be based on precise assessments of the vegetation resources, and these assessments should be regularly updated to detect trends and revise the management plans accordingly.

1.13 Overview and summary

Processes involved in measurements of vegetation characteristics and associated uses with environmental data are given in Figure 1.1. Variability encountered in measurements should not be determined solely after data have been collected. Even casual observations, made visually, reveal sources of variability likely to influence the measures made. The most obvious sources of variation are usually

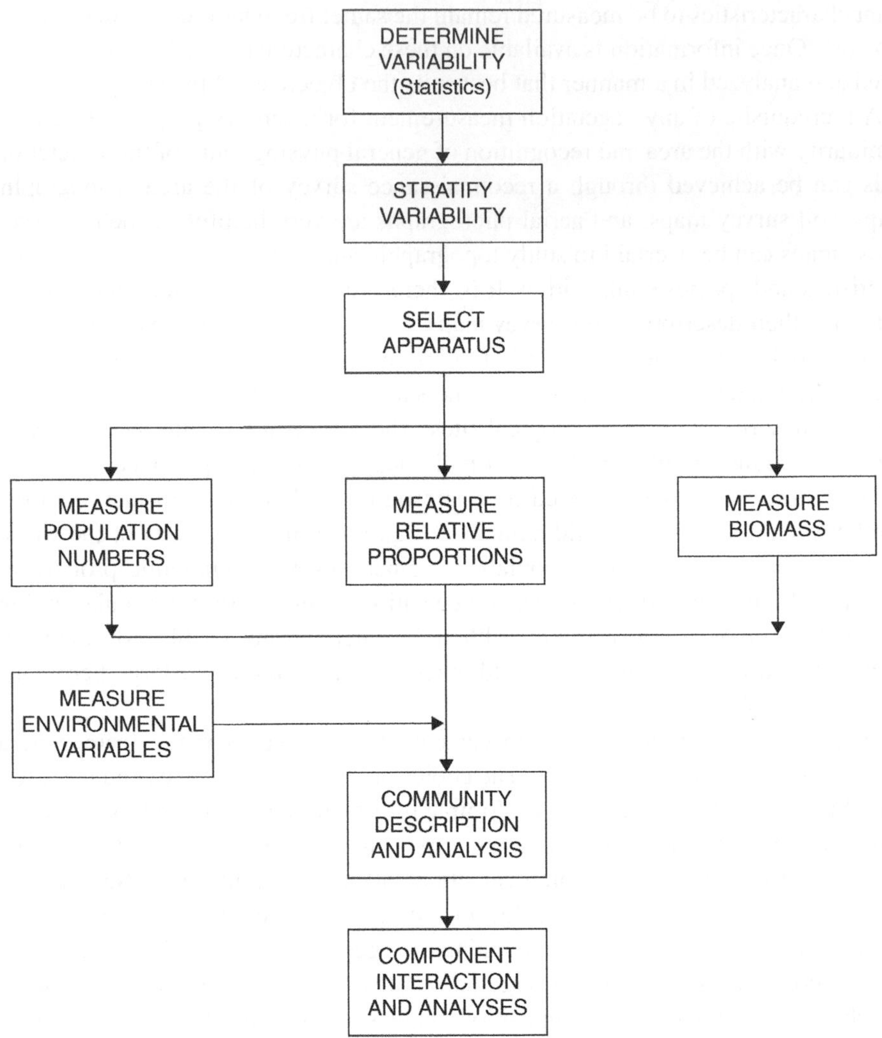

Figure 1.1 A flow diagram of processes involved in vegetation measurement and data analysis.

those influences on vegetation caused by topographic, edaphic, and elevation differences among vegetation types. These same influences are also found in a micro manner within a vegetation type, such as that of individuals of species aggregating into clumps. Thus, one recognizes micro-topographic influences, and micro-relief differences, as well as associated soil differences. All three characteristics of vegetation structure, cover, density, and biomass of species should be used to describe the vegetation of an area. As implied in Figure 1.1, proportions include both cover and frequency. These measures are then used to develop plant community descriptions when environmental data are also available. Basic soil descriptions, along with general meteorological data for the area, are sufficient to provide an analysis of general interactions occurring in vegetation environment of an area. In which case, a vegetation-environmental systems analysis can be used to develop an interpretation for general uses or perturbations that might occur to the vegetation.

1.14 Bibliography

Ashby E. 1935. The quantitative analysis of vegetation. *Ann. Bot.* **49**: 779–802.

Ashby E. 1936. Statistical ecology. *Bot. Rev.* **2**: 221–235.

Bartlett M. 1936. Some examples of statistical methods of research in agriculture and applied biology. *Suppl. J. Roy. Stat. Soc.* **4**: 137–183.

Beasom S. and Haucke H. 1975. A comparison of four distance sampling techniques in South Texas live oak mottes. *J. Range Manage.* **28**: 142–144.

Blackman G. 1935. A study by statistical methods of the distribution of species in grassland associations. *Ann. Bot.* **49**: 749–777.

Bonham C. 1983. Field methods for plant resources inventories. In Conant F., Rogers P., Baumgardner M. *et al.* (eds) *Resource Inventory and Baseline Study Methods for Developing Countries.* AAAS Publication: Washington, DC.

Bonham C. and Clark D. 2005. Quantification of plant cover estimates. *Grassl. Sci.* **51**: 129–137.

Bonham C. and Reich R. 2009. Influences of transect relocation errors on line-point estimates of plant cover. *Plant Ecol.* **204**: 173–179.

Cain S. A. 1934. Studies on virgin hardwood forest. II. A comparison of quadrat sizes in a quantitative phytosociological study of Nash's Woods, Posey County, Indiana. *Am. Midi. Nat.* **15**: 529–566.

Capinera J. L. (Ed.) 1987. *Integrated Pest Management on Rangeland: A Shortgrass Perspective.* Westview Press: Boulder, CO and London, UK.

Clements F. E. 1905. *Research Methods in Ecology.* The University Publishing Company, Lincoln, NE.

Gleason H. A. 1920. Some applications of the quadrat method. *Bull. Torrey Bot. Cl.* **47**: 21–33.

Gleason H. A. 1922. On the relations between species and area. *Ecology* **3**: 158–162.

Gleason H. A. 1925. Species and area. *Ecology* **6**: 66–74.

Green R. H. 1966. Measurement of non-randomness in spatial distributions. *Res. Popul. Ecol. (Kyoto)* **8**(1): 1–7.

Greig-Smith P. 1983. *Quantitative Plant Ecology*, 3rd edn. University of California Press, Berkeley, CA.

Hanson H. C. 1934. A comparison of methods of botanical analysis of the native prairie in western North Dakota. *J. Agric. Res.* **49**: 815–842.

Kylin H. 1926. Über Begriffsbildung and Statistik in der Pflanzen-Soziologie. *Bot. Notiser.* **2**: 81–180.

Odum E. P. and Barrett G. W. 2005. *Fundamentals of Ecology*, 5th edn. Thomson Brooks/Cole: Belmont, CA.

Pipkin F. M. and Ritter R. C. 1983. Precision measurements and fundamental constants. *Science* **219**: 913–921.

Raunkiaer C. 1912. Measuring apparatus for statistical investigations of plant formations. *Bot. Tidsskr.* **33**: 45–48.

Raunkiaer C. 1934. *The Life-Forms of Plants and Statistical Plant Geography*. The collected papers of C. Raunkiaer, translated into English by H. G. Carter, A. G. Fansley, and Miss Fausboll. Clarendon: Oxford.

Svedberg T. 1922. Ett bidrag till de statistika metodermas anvandning inom vaxtbiologien. *Svensk. Bot. Tidsskr.* **16**: 1–8.

Thompson S. K. 1992. *Sampling*. John Wiley & Sons, Inc. New York.

2

Sampling units for measurements

A sampling unit is defined as the basic unit upon which a measurement is taken (Cochran 1977, Thompson 1992). In vegetation measurements the sample unit is most often associated with an area of ground, such as a quadrat, plot, line transect, and so forth, from which a measurement can be obtained on number, ground cover, or biomass of plants. Individual plants are occasionally used as the sample unit. Collectively, the units are referred to as sampling units and should always provide vegetation measurements in terms of an area of ground surface.

Various kinds of equipment and associated methods for their use to obtain measures of vegetation characteristics are presented. Details are presented more thoroughly in the individual chapters on the four measures of vegetation: frequency, cover, density, and biomass. Over time, modifications have been made to both equipment and methods to obtain more efficient estimates of a given measure. Historically, most pieces of equipment were developed for a specific purpose but statistical considerations were not included. As statistical procedures became more important in plant ecological studies, plot and line modifications proliferated. Not all modifications provided unbiased estimates of vegetation measures. Yet, some methods and equipment that yield biased data are still among the most popular in use by vegetation ecologists. Specific uses, with examples, are given in chapters on measures of frequency, cover, density, and biomass.

Variability of a measure is an important consideration when a measurement and associated equipment are selected. For instance, a rectangular-shaped plot will allow more variation to occur within plots than among plots in which the long axis is parallel to the direction of variation for the measure or if spatial scales of the species distribution are less than the plot dimensions.

All equipment used to measure vegetation characteristics has both advantages and disadvantages. No single piece of equipment will provide the greatest efficiency in terms of statistics, economics, and ease of use by the observer. Studies of equipment efficiency in obtaining a measure are often made by statistical procedures such as analysis of variance. In which case, only those studies that include

Measurements for Terrestrial Vegetation, Second Edition. Charles D. Bonham.

equal sample areas are valid. For example, if two plot areas of 0.5 m^2 and 1.0 m^2 are used to obtain estimates of biomass and a sample of 50 is used for the 1.0 m^2 plot, then the sample size of the 0.5 m^2 plot will 100. Then the sample area covered for each plot equals 100 m^2 area.

2.1 Frequency

The sampling unit area of ground enclosed by a plot can be any size or even as large as a plant community, but most likely a plot to determine frequency is often small (for example: 0.1 m^2 or 1 m^2). Plot sizes have been used that ranged from 0.1 m^2 for a moss layer found in plant communities up to and including a plot of 100 m^2 for a forested plant community. Others have used a transparent grid plate that had 100 points arranged in a lattice with 1 cm grid spacing (10 × 10 cm) to estimate frequency. The area of a quadrat used to obtain frequency or other measurement data determines the smallest scale of pattern at which a species occurs (Grieg-Smith 1983). A larger scale of pattern size can be obtained from one-half the average distance (from center to center) among quadrats. Interpretation of results should always include effects of the sampling unit on the scale at which the data applies.

Few large-scale studies have been made using the frequency measure. An example is one by Bergfur *et al.* (2004) who used 0.5 m^2 plots to collect both frequency and cover data to compare species changes within plant communities in Sweden. One of the most complete studies of plot sizes and shapes was made by the United States Forest Service (USA). They recommended that frequency data be collected from quadrat sizes that range from 5 × 5 cm up to and including one that is 100 × 100 cm to obtain rooted frequency in plant communities. These plots can be nested to obtain proper sizes for various plant species.

2.2 Cover

The first known method used to measure vegetation cover was a visual estimate of species cover by category. Later, the contact of a point on a vegetation part was used and continued for over eight decades to measure cover. Individual points are used, as well as several points grouped into clusters. Individual points are represented by tips of sharpened pencils, a length of stiff wire sharpened to a point, cross-hairs in telescopes, or by a laser beam. Areas bounded by plots are probably used more extensively than points to obtain estimates of cover by species, total vegetation, litter, rocks, and bare ground. All kinds of modifications in the uses of points and plot areas have been made to obtain cover estimates in less field time. Only the most frequently used equipment and their respective modifications are presented in Chapter 6.

2.2.1 Points

Points are assumed to be the most objective way to estimate cover. That is, error is minimal for personal bias when a point is used. Either the point contacts a part of the plant, or it does not. The point, or a collection of points, is universally used for cover. Points provide for rapid data acquisition and, thus, are often thought to be the most economical to use. However, errors can occur from other sources, such as movement of plants by wind or improper lowering of the pins by the observer. It must be remembered that a "point" has area that varies with the method used to attain the "point". This becomes important in several ways, as explained in Chapters 4 and 5.

2.2.2 Point frames

Individual points are seldom used to obtain vegetation cover estimates. Rather, points are grouped into clusters. One such method of grouping is called a point frame (Figure 2.1). Usually 10 pins make up the group, but any number can be used. If 10 pins are used, percentage cover can be estimated only within 10% intervals, since each pin is one-tenth of the total contacts possible. Twenty points within a frame or line provide estimates to be within 5% intervals. Pin spacing is arbitrary and can be made to suit the growth pattern of the species. However, points should not be so close in spacing that all 10 points record contacts on a sod-forming

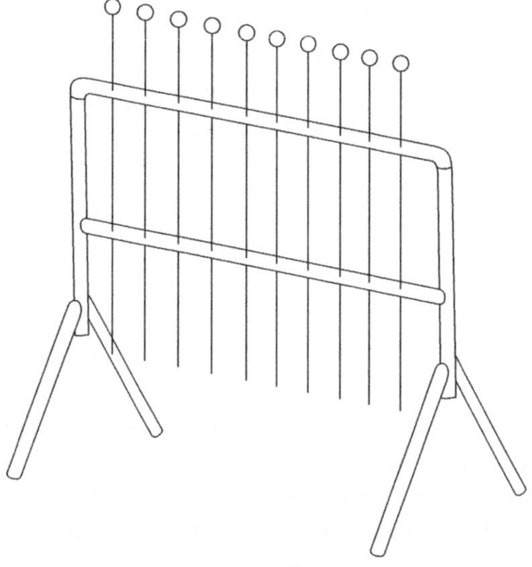

Figure 2.1 Point frame with 10 pins.

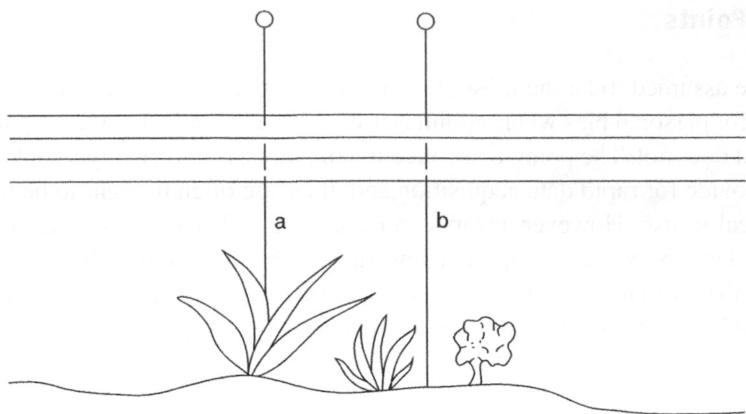

Figure 2.2 Pin a, first hit on plant; pin b, first hit on ground.

grass or that all points make ground contacts between the clumps of the grasses. Vegetation patterns, and in particular species patterns, can strongly influence plant cover results.

Figure 2.2 illustrates a pin where only the first contact is recorded. The first contact was made on a plant leaf by pin *a* and additional contacts would be recorded if the pin were to be pushed further toward ground level. The first contact encountered by pin *b* is the ground itself. First contacts are often recorded so that total vegetation coverage of ground can be estimated. Multiple contacts for a given pin are used for leaf area of species and sometimes for species composition estimates. If an individual pin is selected as a sampling process, then only one pin is used at an observation location. Individual pins can be inclined by either using a single pin in the frame or by construction of a frame to hold only one pin. Rubber bands can be placed around the pin and the frame to create tension so that the pin remains in the raised position until needed.

2.2.3 Point line

A pin placed at any distance along a tape is considered as an extended point frame for point sampling. In this case, a pin is not always necessary because the mark of distance on a tape can be used as a point (Figure 2.3c). If the tape is placed at the vegetation height, then the latter point is easily read as a contact or a no-contact. Otherwise, a pin of some length needs to be lowered from the tape mark until contact is made either with vegetation or the ground. The pin should be lowered at a constant angle or bias will occur. In this case, a frame to steady the pin should be used at each line point. Spacing between points along the line affects results.

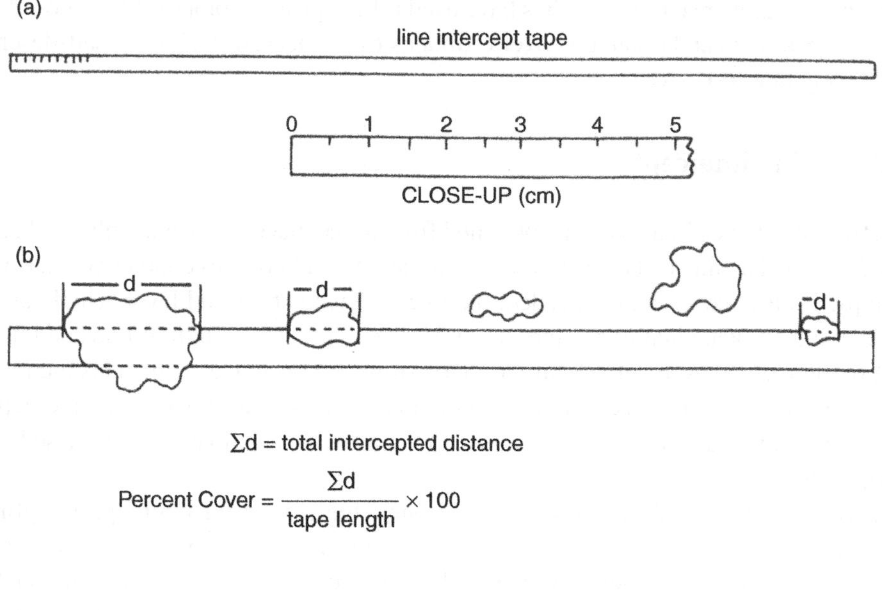

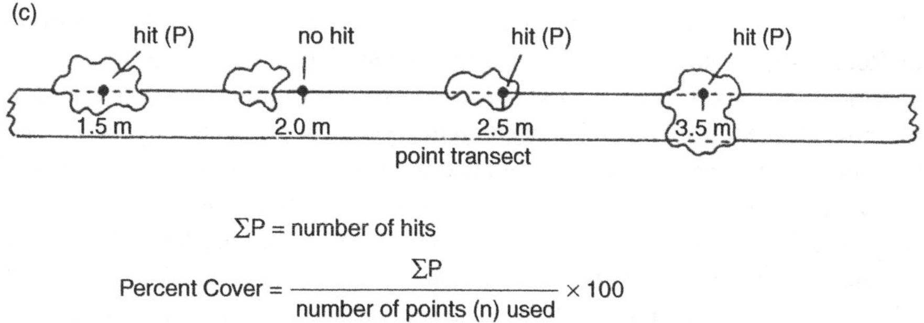

ΣP = number of hits

$$\text{Percent Cover} = \frac{\Sigma P}{\text{number of points (n) used}} \times 100$$

Figure 2.3 Line intercept tape (a), intercept on plants (b), and point transect (c).

2.2.4 Grid quadrats

Grid quadrats have not been used extensively to obtain measures of vegetation cover. These quadrats can be used to estimate frequency at very small scales and cover to the nearest 1%. The assumption made is to have sub-areas in a grid quadrat, each of which is small enough to provide a total quadrat estimate that is precise, reliable, and repeatable. Plant ecologists have used a range of sizes for the primary quadrat. For example, a small transparent grid plate that has 100 grid points arranged in a lattice with 1 cm grid spacing (10 × 10 cm) has been used to estimate cover and frequency in grassland vegetation in Japan and The People's Republic of China (PRC). For this quadrat, one grid point represents 1 cm^2 of cover for the

100 cm^2 quadrat area. Other studies have used a 1 m^2 primary plot gridded into subareas of 5 × 5 cm and obtained cover estimates to the nearest 0.25% for salt-desert shrub vegetation (USA).

2.2.5 Line intercept

Vegetation cover estimates can be obtained from a line placed to contact plants (Figure 2.3a, b). The line or tape is stretched taut at a fixed height to contact vegetation canopy. If basal cover area is needed, the line is placed at ground level. The length of each intercepted plant part is measured (Figure 2.3b). The length of line and total length intercepted by vegetation are used to estimate percentage cover. This method is also used to estimate species composition, that is, the percentage of total intercept made by individual species. Some sources of bias of this estimate are discussed in Chapter 6.

Tapes are often used for lines, and only one edge is read for intercept. Lengths vary, but a range of 6 m to a 30 m length is used to measure vegetation cover. For general inventories, other lengths of 50 and 100 m are used for specific reasons, such as monitoring changes in measures of cover. Locations of the lines are permanently marked by steel stakes so that monitoring can be conducted over a long period of time. Recent studies have shown that these lines cannot be relocated in such a way as to provide unbiased estimates in cover changes over the transect length or over time.

Lengths of line transects used depend on the type of vegetation. In general, cover in herbaceous communities can be estimated with short lines (less than 50 m), while long lines (50 m or greater) should be used in some shrub and tree communities. A transect line does not have to follow a straight line. A few studies have shown that a 90° angle can be used to obtain cover estimates in certain forests. For example, the line length can be divided into half, and the second half continues at a 90° angle to the first half. The proportion of the line (where the line makes a 90° angle) is not restricted to one-half of the line length.

2.2.6 Areas

The various plot sizes and shapes have been used to estimate cover by the use of scales of 1-10, 11-20, and so on to represent cover values that occur within those ranges. Then individual ranges are assigned numerical values such as 1, 2, 3, . . . , 10. Other intervals are used to portray various levels of precision. A plot size and shape of 20 × 50 cm is widely used among vegetation ecologists working in herbaceous communities such as grasslands and in understory vegetation of forests. This plot size and shape is illustrated in Figure 2.4 with two scales, 20% increments and a 1% or 5% coverage scale. The various scales used can be created by subdivision of the plot into percentage intervals representing each scale (1%, 5%, and 20% for

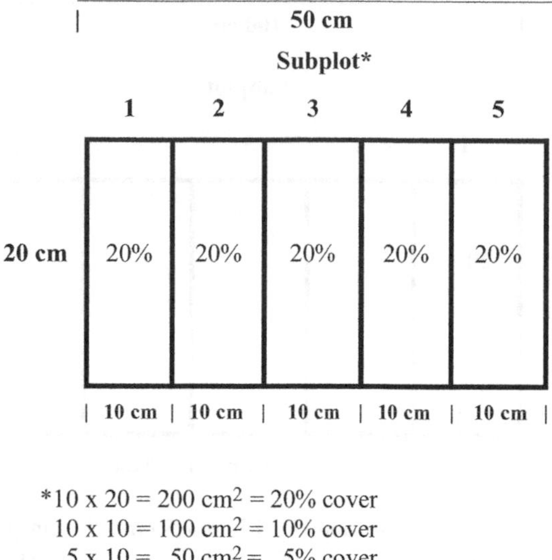

Figure 2.4 Plot used for visual cover estimates. (Note: diagram is not drawn to scale.)

example). Precise divisions can be made by a wire or string connected to the frame to form the 20% or other intervals within the plot.

Care should be taken in the formation of subplot areas so that each of these areas represents a given % value. Each subdivision should divide the plot area into a whole number of units with the subdivision area. For example, a small area for a subdivision that equals 5% of the plot area corresponds to a cover value of 5% for that plot. When the small area is completely covered by plant material it will provide an area that is 5% (5 cm × 10 cm) of the total plot area of 1000 cm^2. A small plot area that is equal to 1% (2 × 5 cm) of this total area provides estimates of cover to the nearest 1% level.

A large rectangular plot of 0.5 m × 1 m can be used to estimate cover to the nearest 1% level by subdividing this plot area into smaller areas (Figure 2.5). Then each sub-plot area of 5 cm × 10 cm represents 1% area of the 20 cm × 50 cm plot above that becomes 20% (1000 cm^2) of the large plot area (5000 cm^2). It is a matter of summing cover estimates to obtain an estimate of total cover the large plot area. Data summaries are described in Chapter 6.

2.2.7 Plotless units

Plots without fixed boundaries that are used to obtain a measure of vegetation cover are called "plotless." This designation is used because the standard method to

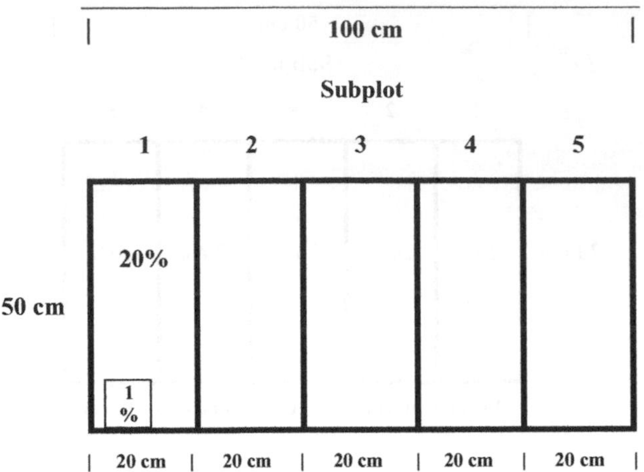

Figure 2.5 A 50 cm × 1.0 cm frame subdivided into a 5 cm × 10 cm plot placed within an individual 20% subplot covers an area of 1% of the whole frame area. (Diagram not drawn to scale).

measure cover depends on an area defined by a fixed boundary. Plotless techniques employ plots of measure which have imaginary and variable boundaries of some fashion. The best known piece of equipment used in a plotless way is probably be the angle gauge (Figure 2.6).

The concept and geometric basis for the angle gauge were developed for use in forest inventory by W. Bitterlich of Germany in 1948. The gauge is variously referred to, but "variable-radius" sampling is a common reference, along with "Bitterlich stick." However, there are so many variations in the dimensions of length and the associated crossbar that the latter designation should be dropped unless the original linear dimensions are used. The angle gauge is based on a small circle (a plant canopy) that occupies 1% of the area of a larger circle if the radius of the large circle is k times the diameter of the small circle. If

$$k = \frac{\text{length of stick}}{\text{length of crossbar}} \tag{2.1}$$

then calculations for the gauge shown in Figure 2.6 are

$$k = \frac{76.2\ \text{cm}}{15.2\ \text{cm}} = 5$$

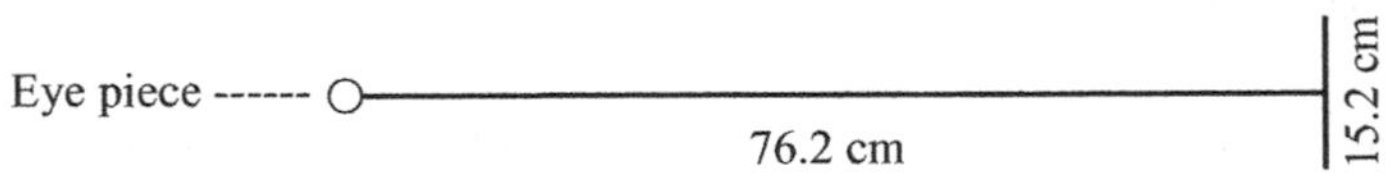

Eye piece ------

76.2 cm

15.2 cm

Figure 2.6 A simple angle gauge stick.

Then, this particular gauge will detect a shrub canopy (or grass, forb, or tree) that is not more than five times its own diameter away from the sample point. Furthermore, that shrub will occupy 1% of the total sample area if a 360° search (to form a circle) is made from the sample point. A count of shrubs that meet this condition will equal the percentage of total area covered by shrubs at that sample point. To obtain percentage cover by species, tally individuals by species as counts are made.

One modification is to place the crossbar so that it can rotate a shaft placed through the main stick. This modification can overcome the effect of slope. Of course, no rotation of the bar is needed if the entire stick can be rotated clockwise to attain the necessary parallelism. The proposed modification is not needed if no peephole or sight is mounted. Then it is simple to just rotate the whole stick. Figure 2.6 illustrates a plant canopy that is more than k times (not counted) its diameter from a sample point and a plant is not at least k times (counted) its diameter from the sample point. Figure 2.7 displays plant canopies to be counted on the basis of

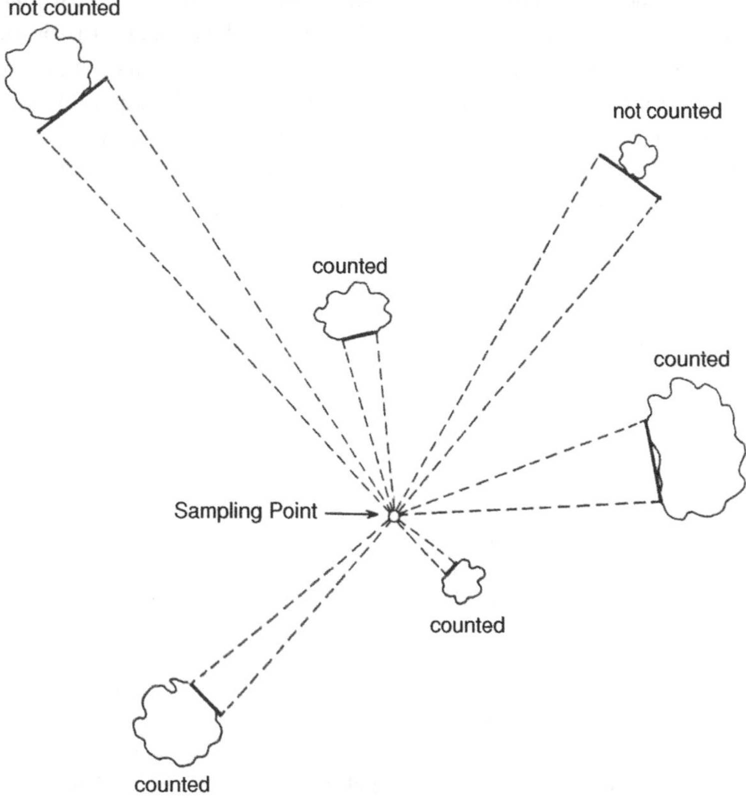

Figure 2.7 Plants considered as within the imaginary plot formed by an angle gauge. (Republished with permission of Society of American Foresters, from Plotless timber estimates-new, fast, easy. Grosenbaugh, L. R. 1952. *J. Forestry* 50: 32–37; permission conveyed through Copyright Clearance Center, Inc.)

the apparent overlap of their canopy with respect to the projected crossbar from a sample point.

A variation in length of the stick and length of its crossbar depends on the particular cover value needed. Ratios of 33:1 cm and 50:1 cm will give basal areas of tree trunks in square feet per acre and square meters per hectare, respectively.

Glass prisms or metal angle gauges are used to measure basal area covered by tree boles. The instruments project a fixed angle. If the diameter of a tree bole is larger than the projected angle, the tree is counted. Otherwise it is not. When sighting through the prism the observer will see a section of the bole offset from the remaining part of the bole. If this offset is discontinuous the tree is not counted. Otherwise it is. Distance from the sample point to the tree determines the amount of misalignment of the bole as viewed through the prism. The count is multiplied by a constant to estimate basal area covered. Decisions on trees to be counted are illustrated in Figure 2.8. Prisms are available in various angle sizes depending on the density and size of trees.

Other plotless methods include distance measures that are used to estimate mean area per plant, and subsequently, cover can be estimated by species or for total vegetation. First, density estimates are made using a selected distance measure (from Chapter 7), and for each distance, the plant is measured for diameter. Diameters are used individually to calculate areas of ellipses or circles assumed by plant canopies. Then, an average is obtained for area covered by plant canopies. Density multiplied by the average ground coverage by plant canopies will give the total area covered.

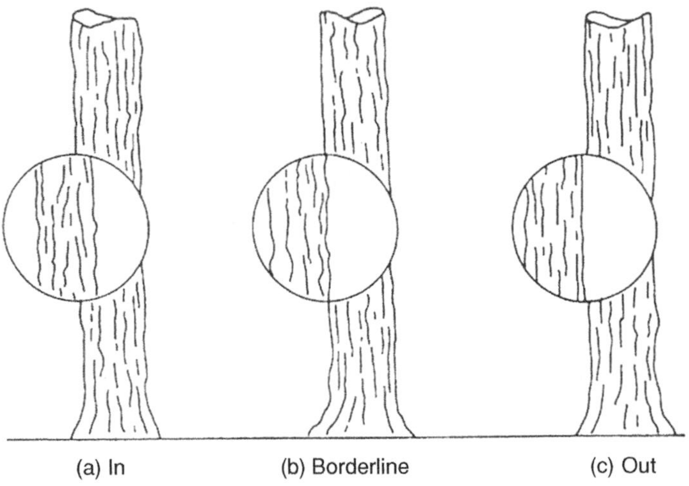

Figure 2.8 Displacement of tree boles viewed through a prism. (Reproduced, with permission, from *Forest Mensuration*. Husch, B., Beers, T. W., and Kershaw, J. R. 2003. © John Wiley & Sons, Inc.

The same plots used for density, such as individuals per hectare, are used to calculate percent cover. For example, consider an area of one hectare in size, then

$$\text{Percentage plant cover} = \frac{(\text{density}) \times (\text{average area of plant})}{10\,000} \times 100 \quad (2.2)$$

Consider several assumptions made to obtain precise estimates of cover from distance and diameter measures. In particular, consider precision: (a) of density estimates, (b) of estimated areas covered by plants, and (c) of the product of the two estimates.

2.3 Density

The majority of plant density measures are made from plot or distance measures. Equipment used as a plot frame to obtain the measure is rather simple. Plot frames are easily made from welded metal and wire if areas are 1 m^2 or less. Otherwise, boundaries are marked in the field with string or pliable wire held in place by metal pegs. Distance measurements require only a tape measure, whether made out of cloth or metal.

2.3.1 Plot size

Plots with boundaries of 1, 4, and 10 m on a side form 1, 16, and 100 m^2 plots, respectively (Figure 2.9), and are used extensively when counts of individual herbs, shrubs, and trees are needed from a forested community. Usually these plots are nested, that is, the 1 m^2 plot is placed in a corner of the 16 m^2 plot, and the latter is placed in a corner of the 100 m^2, and the latter plot is placed in like manner within a 900 m^2 plot. Only herbaceous plants are counted by species in the 1 m^2 area, while shrubs are counted in the total 16 m^2 area, and small trees in the 100 m^2. Large trees can be counted within the largest plot, 900 m^2 that is associated with a 30 × 30 m pixel generated by Landsat TM imagery. This imagery is used in various ways to locate and establish sampling designs, as presented in Chapter 4.

2.3.2 Shape of plots

Shapes of plots include squares, rectangles, and circles to estimate density of plants. Studies on plot shapes might have originated in the early twentieth century when Raunkiaer (1909) began recording frequency of plant species. He did not refer to a specific study. He elected to use a circular plot of 0.1 m^2 to list occurrences of plant species in plant communities. The same could be inferred from a publication

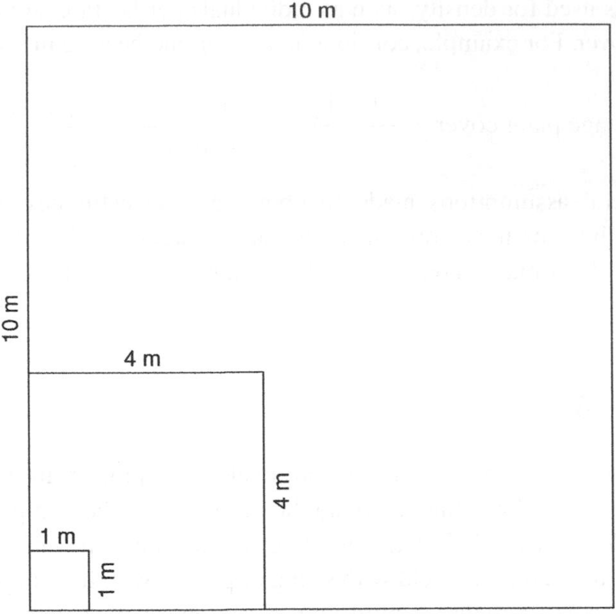

Figure 2.9 Nested plots used to count herbaceous plants, shrubs, and trees, respectively, for 1, 16, and 100 m^2.

by Pound and Clements (1898). They are known to have been among the first to use quadrats to measure vegetation characteristics. They used 5 m^2 frames to count plant species in vegetation types in Nebraska (USA) but they realized that estimation of density was affected by the degree of dispersion of individuals and quadrat size and shape. More recent studies have reported effects of quadrat shape on results of frequency, cover, density, and biomass estimates. Specific results are given in the respective chapters for these measures.

2.3.3 Line transects

Line transects can be used to measure the amount of intercept of plants along a straight line and these intercepts are then used to obtain estimates of density. Figure 2.3b illustrates the measure of intercepted length of a shrub or other plant. These lengths give unbiased estimates of cover, not density. To obtain estimates of density, the intercepted plant is measured for diameter that is interpreted as widths of canopy rather than actual diameter. Alternatively, interception of chords of circles formed by plant canopies also is a basis for obtaining estimates of density from line intercepts. These widths provide a measure of area and counts within that area. Equations to convert these widths into an estimate are given in Chapter 7.

2.3.4 Distance methods

These methods have been used for several decades (beginning *ca.* 1940s) to estimate plant density of trees and shrubs. These methods also have been referred to in literature as "variable-plot" and "plotless" techniques. Several of the techniques use the assumption of an existing circle wherein plants occur. Distance (radius) is measured from various plant-to-plant pairs and distances from randomly located points in the field. In each case, the distance is used to estimate an area associated with plants and, thus, obtain estimates of density. Even estimates for basal areas and canopy foliage for trees have been estimated from these methods. Some distance-based estimates of density, cover, and biomass have been obtained in herbaceous plant communities. Only distance methods that result in unbiased estimates for density are described. These methods can be used when density is either random or non-random in its distribution within the plant community. All other methods yield biased estimates.

2.4 Biomass

Vegetation biomass is most often determined from plots of given areas. Vegetation areas are usually clipped and weighed to yield an estimate of biomass. Plots of various shapes and sizes are used to obtain the measurement. Shapes include circles, rectangles, and squares, while sizes range from a few square centimeters to several square meters. In general, dense herbaceous vegetation makes small plot areas more efficient, while sparse vegetation allows the use of large areas for biomass measurement. Choice of plot size and shape are usually based on past studies conducted within a plant community as that of a present study. Any differences in estimates of biomass from these plot characteristics (i.e., size and shape) can be overcome by sample size and associated sample unit areas covered in the study.

Small areas described by plots may yield biased estimates of biomass. If individual plants of bunchgrasses, for example, are so large as to exclude the occurrence of any other plant within the plot boundary, then biomass of the large plants is overestimated, while the biomass of other species is underestimated. Additionally, small plots have larger area:boundary-length ratios. Table 2.1 gives a comparison of the area:boundary ratios for several plot sizes (area) and three plot shapes. If area is constant (e.g., 1000 cm^2), then plot shape can be changed to increase the amount of area A for a given boundary length B. Note that for a fixed area of 1000 cm^2, the A:B ratio increases from a rectangle, to a square, to a circle (from 7.14 to 8.92). This represents a 25% increase in area per plot of boundary length for the circle compared to the rectangle.

Table 2.1 Boundary length and area relationships as affected by plot size and shape

Shape (cm)	Boundary Length B (cm)	Area A (cm^2)	A:B (cm)
20 × 50 rectangle	140.0	1000	7.14
50 × 100 rectangle	300.0	5000	16.67
50 × 200 rectangle	500.0	10 000	20.00
31.62 × 31.62 square	126.5	1000	7.91
70.71 × 70.71 square	282.8	5000	17.68
100 × 100 square	400.0	10 000	25.00
17.84 (radius) circle	112.1	1000	8.92
39.89 (radius) circle	250.7	5000	19.94
56.42 (radius) circle	354.5	10 000	28.21

Larger areas of plots show less increase in plot area per boundary length. For example, the percentage increase for a circle of 1000 cm^2 compared to an area of 5000 cm^2 is 124%. Yet, a twofold increase in area (5000 cm^2 to 10 000 cm^2) yields an increase of 41% in the amount of area per plot of boundary length. The same increases also hold for the square plots of comparable areas (Table 2.1). On the other hand, the rectangular-shaped plots have a ratio increase of 14 and 67%, respectively, for an increase in area from 1000 to 5000 cm^2 and from 5000 to 10 000 cm^2.

Based on data in Table 2.1, decisions made to include plants that occur very near or on boundaries of plots are minimized for circular plots. This is true only for plot shape comparisons based on constant area for all shapes. Otherwise, an increase in area or size of a plot can be made for any shape until the area:boundary-length ratio is acceptable. For example, a decision is made to accept a plot, rectangular in shape ($50 \times k$ cm), to provide a 10% or less increase in the ratio. Then k is determined to be 180 cm from Figure 2.10. That is, the 10% increase in the ratio is found at a boundary length of 460 cm (area of 9000 cm^2). Then, 460 minus 100 (for two sides of 50 cm each) equal 360 cm, to be divided equally for the other two sides. A rectangle of dimensions 50 × 180 cm gives the desired ratio. Any larger k for the plot will give less than 10% increase in area:boundary-length ratio.

2.4.1 Plot characteristics

Plots commonly used to obtain weight estimates are illustrated in Figure 2.11. All plots cover 0.5 m^2 of area. Boundary lengths and area ratios are given in Table 2.1. More boundary decisions are expected to be made for the rectangle plot because more boundary is involved than with either a square or a circle of the same area. Therefore, measurement error is thought to be less for circular-shaped

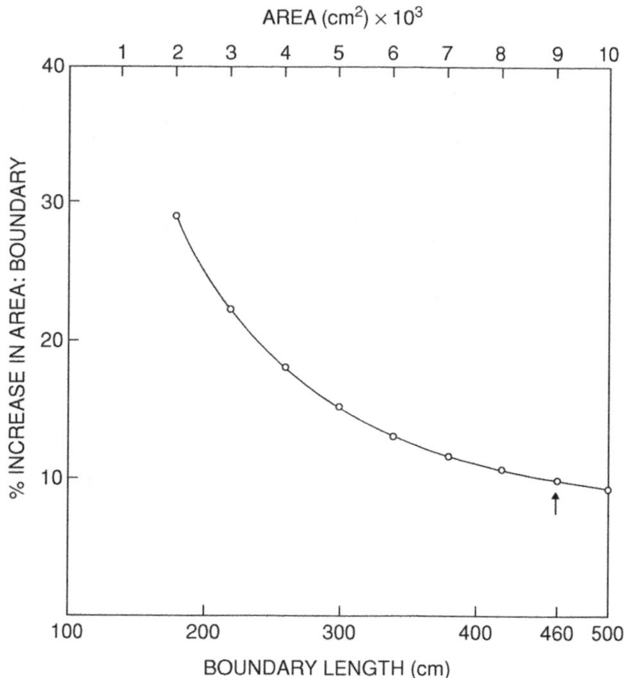

Figure 2.10 Boundary length versus percentage increase in area:boundary ratio.

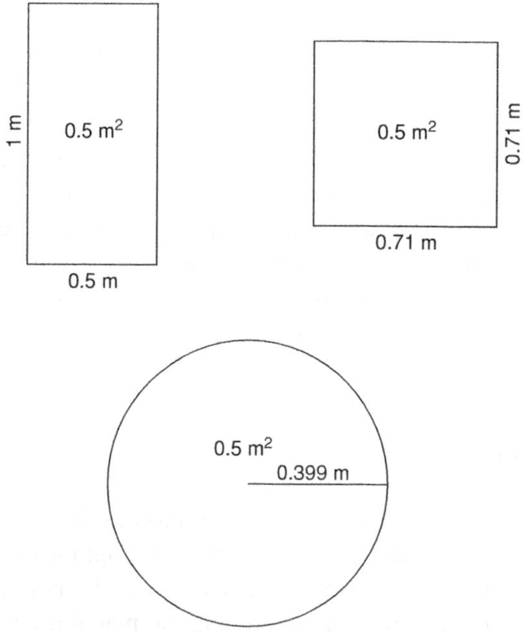

Figure 2.11 Plot dimensions for rectangle, square, and circle of 0.5 m^2.

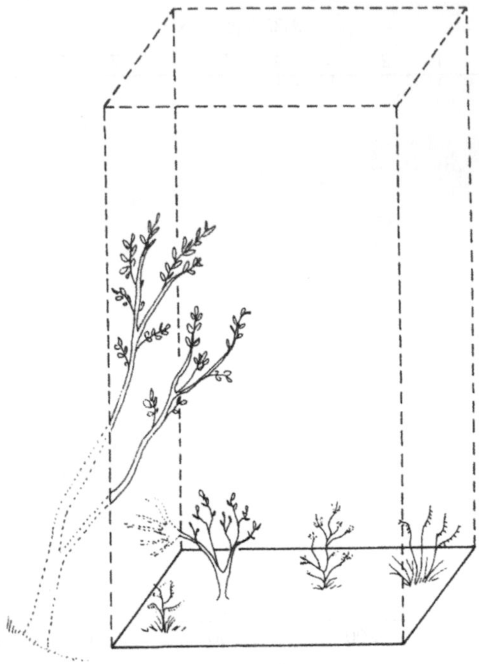

Figure 2.12 Theoretical volume over a plot area. Plants clipped include all material within the volume, not just that rooted within the plot.

plots. In practice, however, the three shapes are just as efficient in terms of sample variability.

All vegetation biomass is contained within a volume of space, not area alone (Figure 2.12). That is, the estimate of biomass is based on the projected material above ground level, not on rooted material only. An imaginary bounded volume must be followed in guessing and/or clipping material to estimate biomass. Figure 2.13 is an example of a plot used to sample for biomass estimates in tall herbaceous vegetation; this plot defines the vertical boundaries of a plot by extending it into the vertical space above (Figure 2.13).

2.4.2 Indirect methods

The weight-estimate method for measurement of biomass in plots has been used for several decades. It is also referred to as double-sampling; meaning that two observations are made from the same plot, visual estimates of amount of plant biomass in the plot and an estimate obtained by clipping the material and weighing it with a balance. The method is cost efficient in the sense that only a small number of

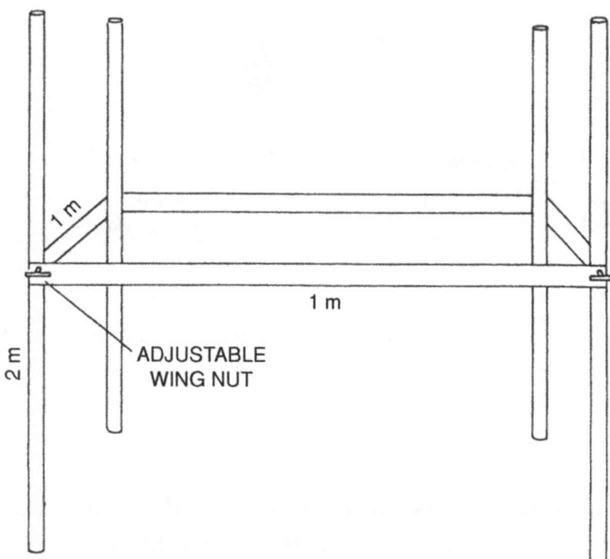

Figure 2.13 Plot used to clip vegetation by height increments.

plots are clipped compared to a larger number of plots that are only estimated using ocular methods. For instance, a ratio is decided before sampling that one out of four plots (1:4) will be both visually estimated and clipped while the 3:4 plots will be visually estimated. Ratios used in sampling for biomass measurement can range from 1:2 for small areas (paddocks of say, 10 ha or less) up to 1:50 or more for large areas (over 100 ha).

An alternative to plots with rigid boundaries, for example, is an area inscribed by use of a string, pipe, or metal rod. A 1 m^2 area is defined with an radius of 0.56 m in length. Any area can be found by a circle made with simple materials (string) used for the given radius. Place the string on a stake that is driven into the ground and inscribe a circle on the ground with a rod sharpened on one end.

2.5 Tree measurements

Equipment specifically used to obtain measures of tree characteristics is numerous, and only a few are presented in this book because several references provide detailed descriptions for all equipment.

2.5.1 Tree heights

Heights of standing trees are usually estimated indirectly by instruments called hypsometers that use geometric or trigonometric principles. Hypsometers based on the

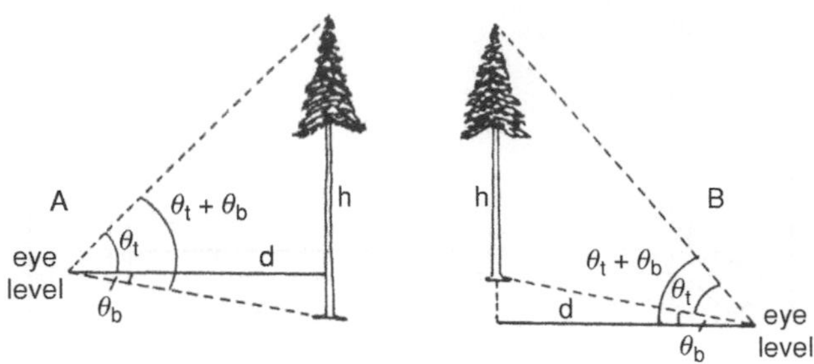

Figure 2.14 Trignometric relations used with hypsometers to estimate tree heights (Reproduced, with permission, from Belyea, H. C. *Forest Measurement*. 1931. © John Wiley & Sons, Inc.)

principles of similar triangles include the Christen and Merrit instruments that are sticks with graduated scales (Husch *et al.* 2003). Some hypsometers require a measurement of the horizontal distance from the observer to the tree. If a slope exists, then corrections for horizontal distance are obtained by trigonometric functions. Some hypsometers require that the distance between the tree and a point be constant, while other hypsometers do not depend on distance to estimate heights.

Hypsometers based on tangents of angles usually provide better estimates of tree heights (Husch *et al.* 2003) (Figure 2.14). Principles for use of area and angles enable the construction of various hypsometers. These principles are employed in the use of the Abney level, the Haga altimeter, the Blume–Leiss altimeter, and the Suunto clinometer (Figure 2.15). These are widely used hypsometers for height and slope measurements. Each type of hypsometer has advantages and disadvantages that depend on the topography and density of the trees. In general, the measurement is obtained from a position where both the top and base of the tree can be seen. Line of sight angles to the top and base of the tree are observed.

Tree height is estimated by application of equations using tangents or sines. Note that if the horizontal line of sight intersects a tree between the base and the top, two estimates are made of the height, and these two estimates are summed to obtain the total height. Should a measure involve a line of site above or below the base because of slope, then tree height is the difference of the two estimates (Belyea 1931).

The use of tangents for angles requires a measure of the horizontal distance from a sample point to the tree. Belyea (1931) gives equations to estimate tree heights when angles are measured. If the eye of the observer is above the level of the base of the tree (i.e., up slope from the tree) (Figure 2.14a), then

$$h = \frac{d[\sin(\theta_t + \theta_b)]}{\cos\theta_t} \tag{2.3}$$

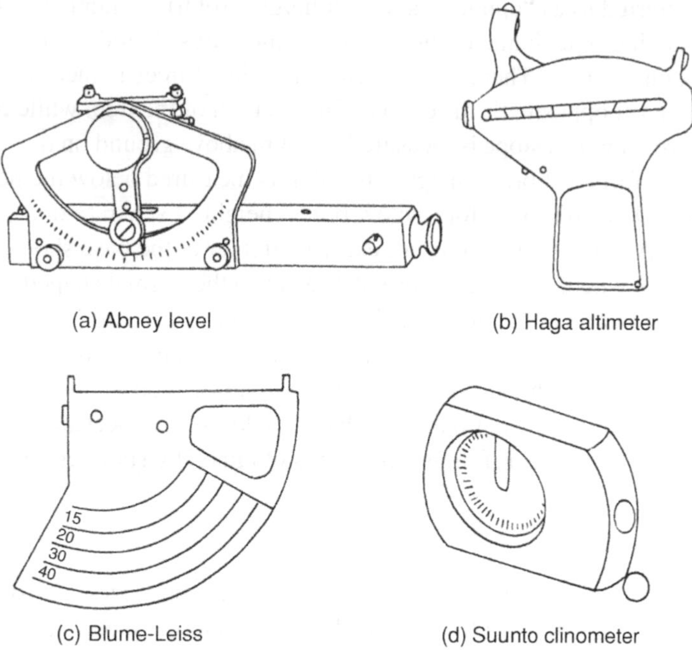

(a) Abney level (b) Haga altimeter

(c) Blume-Leiss (d) Suunto clinometer

Figure 2.15 Commonly used hypsometers (Reproduced, with permission, from *Forest Mensuration*. Husch, B., Beers, T. W., and Kershaw, J. A. 2003. © John Wiley & Sons, Inc.)

On the other hand, if the eye of the observer is below the level of the base of the tree (i.e., down slope) (Figure 2.14b), then d takes on a different position.

$$h = \frac{d[\sin(\theta_t + \theta_b)]}{\cos\theta_t} \tag{2.4}$$

If trees are not erect, then the vertical distance from the ground to the top of the tree is measured at right angles to the direction of the lean. Grosenbaugh (1981) described a hypsometer that rotates to provide estimates for the heights of trees that lean. Any hypsometer may not adequately estimate the height of deciduous trees since exact maximum height may not be seen and is assumed to occur somewhere beyond the first canopy intercept.

2.5.2 Units of measure for tree diameters

Tree diameters are obtained from instruments that range from simple to sophisticated devices. These include calipers, tapes, and optical instruments. A diameter measure of trees is usually recorded at breast height (1.4 m), and is commonly referred to as "diameter at breast height" (dbh). A number of field rules exist for obtaining bole diameter measures. All measures should be obtained at the same height on all trees. Then, the diameter of inclined trees is measured at 1.4 m from ground level opposite the direction of inclination of the tree, while the diameter of trees growing on a slope is measured at 1.4 m above ground on the uphill side. If a tree bole is forked at breast height, then dbh is measured below the point where the bole forks. For boles that fork below breast height, a separate dbh is recorded for each fork. Any tree bole that is not exactly circular at the base should have dbh measured above where the bole is normal relative to the normal shaped bole.

Estimates of tree volume are made from "diameter inside bark" (dib) measures. To obtain such a measure, a small portion of the bark must be removed at breast height to measure bark thickness. "Bark gauges" can be used to minimize injury to the tree (Mesavage 1969). Estimates for bark thickness are measured at points on opposite sides of the bole. It follows that dib is obtained by subtraction.

Tree calipers

Calipers used to estimate diameters of trees are of various sizes and shapes (Figure 2.16). Most calipers use a fixed scaled bar while another piece slides on the bar. The method of obtaining the measurement for diameter depends on the construction of the instrument. To read diameter, an average of two diameter measurements is obtained of the longest and shortest diameter for non-circular boles.

Diameter tape

The diameter of the tree boles can also be measured with a "diameter tape." This tape is like an ordinary measuring tape, but has one side marked in linear units which are converted to a diameter for any given circumference of a circle. The conversion is based on the relationship between the circumference and the diameter of a circle. That is

$$D = \frac{C}{\pi} \tag{2.5}$$

where

D = bole diameter
C = bole circumference
π = 3.14

It follows that a circular shape is assumed for the bole.

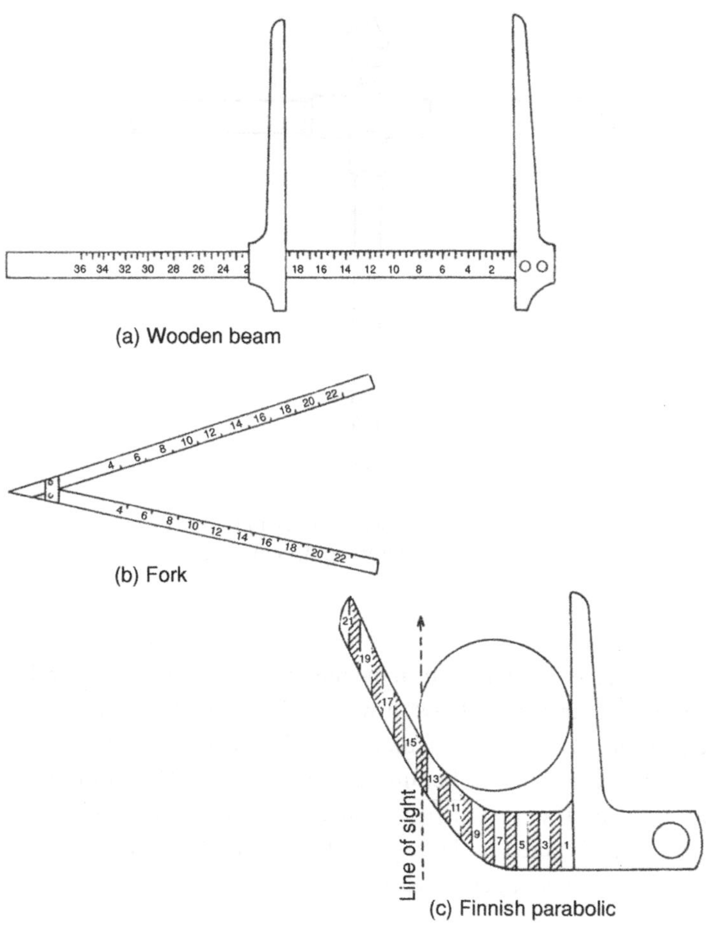

Figure 2.16 Three types of tree calipers (Reproduced, with permission, from *Forest Mensuration*. Husch, B., Miller, C. I., and Beer, T. W. 2002. © John Wiley & Sons, Inc.)

Increment borers

Growth increments in bole diameters are measured by borers. Since the instruments were designed by Swedish foresters, they are often called "Swedish increment borers." A typical increment borer (Figure 2.17) consists of a hollow auger that is drilled into the tree until it intersects the growing center of the tree in a plane perpendicular to the longitudinal axis of the tree. The auger is carefully turned backwards a fraction of a turn to break the wood core, and then the sample core is removed for counting growth rings and measuring the width of each ring.

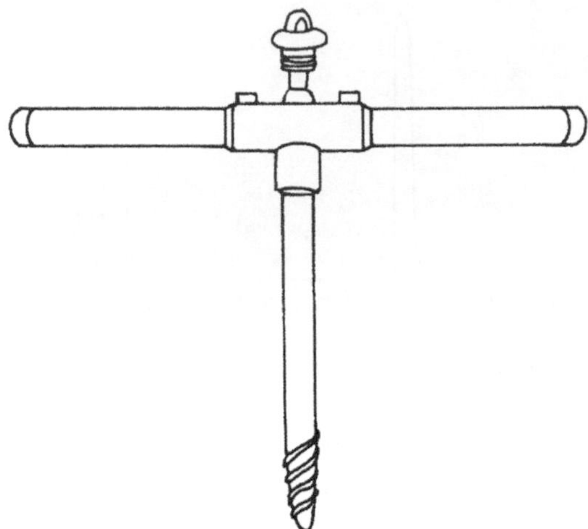

Figure 2.17 Increment borer (Reproduced, with permission, from *Forest Mensuration*. Husch, B., Miller, C. I., and Beer, T. W. 2002. © John Wiley & Sons, Inc.)

The age of the tree is estimated from the number of growth rings. Problems in the estimation occur because some trees grow at a differential rate in various seasons, while some growth is affected by meteorological events (precipitation, etc.). "False rings" are present if drought or other unusual events occur. Only through experience in the study of material obtained from cores will interpretation be accurate. Such experience must be obtained in the region of interest. Otherwise, increment borers are not suitable to estimate ages of trees.

2.5.3 Tree crown cover

The "Moosehorn" is used to estimate the percentage of crown closure or crown cover for stands of trees. Although it was originally developed to estimate volume of timber, Garrison (1949) modified the instrument to obtain crown closure estimates (Figure 2.18). The dot template shown is transparent and the spots are red so that they contrast with live foliage. The instrument is a wedge-shaped box with a sighting device and is mounted on a standard surveyor's staff. Mirrors are used to create a vertical projection of the canopy via reflection into the box and onto the dot template. The proper orientation is obtained when images of a marker fall on the center line of a mirror. Line of sight is attained when the cross-hair is centered inside the peep-sight ring. Cover is estimated by counting dots on the template covered by the overstory.

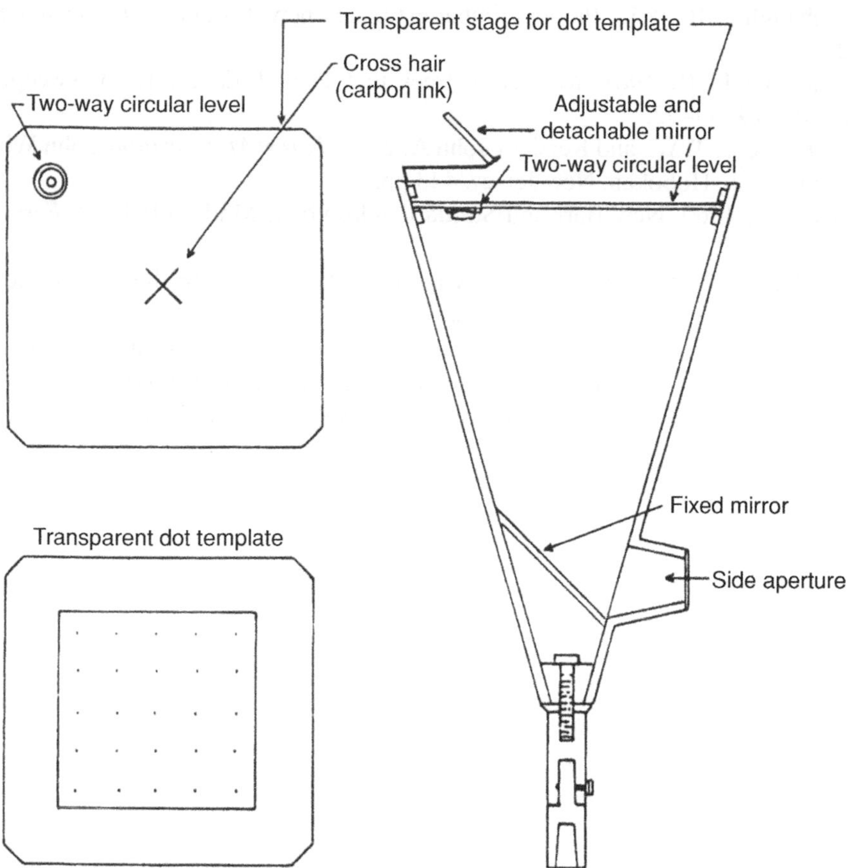

Figure 2.18 A "moosehorn" to estimate crown closure. (Republished with permission of Society of American Foresters, from uses and modifications for the "moosehorn" crown closure estimator. Garrison, G. A. 1949. *J. Forestry*, **47**: 733–735; permission conveyed through Copyright Clearance Center, Inc.)

2.6 Bibliography

Belyea H. C. 1931. *Forest Measurement*. John Wiley & Sons, Inc.: New York.

Bergfur J., Carlsson A., and Milberg P. 2004. Phenological changes within a growth season in two semi-natural pastures in southern Sweden. *Ann. Bot. Fennici* **41**: 15–25.

Cochran W. G. 1977. *Sampling Techniques*. John Wiley & Sons, Inc.: New York, 428 pp.

Garrison G. A. 1949. Uses and modifications for the "moosehorn" crown closure estimator. *J Forestry* **47**: 733–735.

Grosenbaugh L. R. 1952. Plotless timber estimates-new, fast, easy. *J. Forestry* **50**: 32–37.

Grosenbaugh L. R. 1981. Measuring trees that lean, fork, crook, or sweep. *J. Forestry* **79**: 89–92.

Husch B., Beers T. W., and Kershaw John A. 2003. *Forest Mensuration*. John Wiley & Sons, Inc., Hoboken, New Jersey. 443 pp.

Mesavage C. 1969. New Barr and Stroud dendrometer, Model FP 15. *J. Forestry* **67**: 40–41.

Pound P. and Clements F. E. 1898. A method of determining the abundance of secondary species. *Minn. Bot. Stud.* **2**: 19–24.

Raunkiaer C. 1909–1910. Formationsundersogelse og formationsstatistik. *Botanisk Tidsskrift* **30**: 20–132. (English translation Raunkiaer 1934 below).

Thompson S. K. 1992. *Sampling*. John Wiley & Sons, Inc.: New York.

3

Statistical concepts for field sampling

The purpose of this chapter is to introduce fundamental statistical concepts and to develop a basis for their use in field vegetation measurements. This chapter does not address statistics as a discipline in detail, as the subject is presented in many books. Instead, emphasis is placed on the underlying principles of statistics, an understanding of which is important to obtain and assess vegetation measurements in an objective manner. Statistical procedures are, of course, very effective ways to summarize and evaluate data.

The statistical approach to making field measurements provides an objective method of obtaining and evaluating vegetation data. Therefore, any methodology that does not incorporate statistical criteria should be used with caution in vegetation measurements. Individual chapters on vegetation measurements will emphasize principles to assist in the selection of a methodology that meets minimum statistical criteria.

Application of statistics in field measurements of vegetation characteristics may be divided into three steps:

1. *Design of the sampling process*. An efficient sampling plan will lead to a reduction in unnecessary data collection. Any sample design should include, beforehand, the kind of statistical analysis to be conducted on the vegetation data collected. It may not always be possible to specify the exact details of all analyses. But at least some general form of analysis, such as analysis of variance or a *t*-test, should be in mind before any measurement is made.

2. *Reduction of data for estimation purposes*. Data reduction usually results in a description of a characteristic of the sample. Such a description is referred to as a "statistic" and may include such estimates as the mean (average) or variance (spread) of the sample data.

Measurements for Terrestrial Vegetation, Second Edition. Charles D. Bonham.

3. *Examination of significance*. Information may be gained if hypotheses are tested. For example, we may question whether the population mean differs statistically from one vegetation type to another. Such a mean value might be total standing biomass or any other vegetation characteristic. Significance, or the lack thereof, should be determined from a randomly drawn sample of the vegetation characteristic to be measured and by testing an appropriately constructed hypothesis about the sampled population.

3.1 Characterization of data

Vegetation data can be presented in one of several ways, including a list, class intervals, or a summary form such as an average. A list of data values is more easily read and interpreted if data are arranged from the smallest value to the largest value or *vice versa*. Class intervals may be used to classify data according to a given interval width. Then, the number of times a value occurs in each class interval can be used to calculate the proportion of values that occur within each class interval. This proportion is referred to as the "relative frequency" of the data. Relative frequencies of data values indicate probabilities of occurrences for the interval values, and this distribution is referred to as a "frequency distribution for the data values." For example, the probability of occurrence, or frequency distribution, of a particular face of a die is 1/6 and happens to be the same for all faces. These probabilities (1/6, 1/6, 1/6, 1/6, 1/6, 1/6) indicate the frequency distribution of each numerical value on the six faces of a die, that is, the numerical values 1 through 6.

The biomass data in Table 3.1 is a population of possible values in grams per square meter, which has a frequency distribution. This distribution is presented as a histogram (Figure 3.1). If the same data in Table 3.1 are considered as a sub-set of a larger population, the frequency distribution shown in Figure 3.1 will be the frequency distribution of a sample or of a sub-population.

The histogram in Figure 3.1 describes the central location and dispersion parameters of the frequency distribution. The location of these data occurs somewhere between the values of 12 and 72; most observation values occur between 36 and 48. The location of data can be summarized by various measures of central tendency and include the arithmetic mean, the mode, and the median, which are discussed later.

The spread or dispersion of the data is measured by the range in data values, the quartile range, the mean deviation of individual values from the mean, and the standard deviation. These measures of location and dispersion are used to describe sample data or the entire population of values. When these measures are applied to populations, they are referred to as "parameters," but are defined as "statistics" when applied to sample data or a sub-population data set.

Table 3.1 Grams of green plant material per square meter[a]

(1)	(11)	(21)	(31)	(41)
28	42[b]	38	53	32
(2)	(12)	(22)	(32)	(42)
39	50[b]	49	45	57
(3)	(13)	(23)	(33)	(43)
22[b]	46[b]	54	43	37
(4)	(14)	(24)	(34)	(44)
39	44	45[b]	37	46
(5)	(15)	(25)	(35)	(45)
48	12	44	71	45[b]
(6)	(16)	(26)	(36)	(46)
61	29[b]	31	35	26
(7)	(17)	(27)	(37)	(47)
58	41	37	30	49
(8)	(18)	(28)	(38)	(48)
52	66	43	26	39[b]
(9)	(19)	(29)	(39)	(49)
45	40	59	39	38
(10)	(20)	(30)	(40)	(50)
48[b]	33	39	14	27[b]

[a]Assuming that a plot of 5 × 10 m is the population. Numbers in parentheses indicate plot number. A sample of 10 plots is drawn at random to estimate population parameters.
[b]Drawn at random to be measured.

3.1.1 Measures of data location

The center of a data distribution, like that shown in Figure 3.1, can be measured in several ways. The center of a distribution of population values is referred to as a "location parameter," which describes the central tendency and is calculated by summing data values together and then dividing this sum by the number of values used in the summation. The process in steps is straightforward. The sample data (Table 3.1) are used to estimate the "arithmetic mean" and

$$\sum_{i=1}^{n} X_i = 393 \text{ g}/10 \text{ m}^2 \tag{3.1}$$

The Greek letter Σ represents addition or summation, and the index $i = 1$ to n indicates that the summation extends over all data values, where n represents the number of data values for the variable X and $n = 10$ for the example. The arithmetic mean of a population is represented by the Greek letter μ, and $\bar{X}$ represents the arithmetic mean of a sample from the population. It is more convenient to let N denote the number of sample units that occur in the population and n the number of

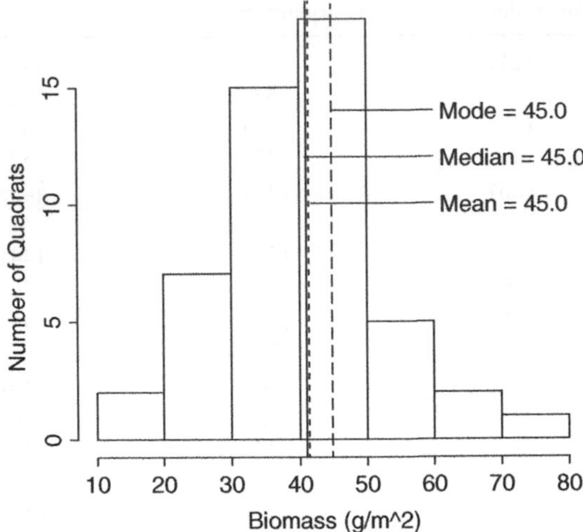

Class Size	Frequency	Probability (%)
10–19	2	5.0
20–29	7	17.5
30–39	15	37.5
40–49	18	45.0
50–59	5	12.5
60–69	2	5.0
70–79	1	2.5
Total	50	100

Figure 3.1 Histogram of biomass from 50 m^2.

sample units in a sample. In the example (Table 3.1), $N = 50$ and $n = 10$. Therefore, from Table 3.1, the sample mean is

$$\bar{X} = \frac{\sum_{i=1}^{n} X_i}{n} = \frac{393\,\text{g/m}^2}{10} = 39.3\ \text{g/m}^2 \tag{3.2}$$

and the population mean is

$$\mu = \frac{\sum_{i=1}^{N} X_i}{N} = \frac{2060\,\text{g/}50\,\text{m}^2}{50} = 41.2\,\text{g/m}^2 \tag{3.3}$$

Note that $\bar{X} = 39.3\ \text{g/m}^2$ is an unbiased estimate of $\mu = 41.2\,\text{g/m}^2$ if the sample were obtained by random sampling methods.

The second measure of central tendency is called the "median," which is the middle value of a data set if n or N is odd, or it is the average of the two middle items if n or N is even. The median value m for the sample previously used is determined to be

$$m = \frac{42 + 45}{2} = 43.5 \tag{3.4}$$

This value is obtained by first ordering the sample data from low to high values (g/m^2). So, the order is

$$22, 27, 29, 39, 42, 45, 45, 46, 48, 50$$

Since $n = 10$, an even number, the two middle values are 42 and 45.

For the population of Table 3.1, the median is

$$M = \frac{41 + 42}{2} = 41.5 \tag{3.5}$$

The population median could also be found from a vertical line that exactly bisects the area covered by the histogram in Figure 3.1.

The third measure of central tendency is the "mode," which is the value that occurs most often in the frequency distribution. The mode for the population is a value somewhere between 36 and 48 g/m^2 (Figure 3.1). This range is the modal class. From values given in Table 3.1, the modal value is 45 g/m^2, because this value occurs most often. Both population and sample distributions have modal values.

We have considered general methods for data descriptions by use of a graphical display and by calculation of the arithmetic mean, mode, and median, which are all measures of central tendency. Each of these measures takes on a value that describes or represents the sample data set or the entire population. Because all three measures indicate central tendency or location, the selection of which one to use depends on the following:

1. What is the need of a typical or representative value as dictated by the problem? Is an absolute or a relative value needed, or is a middle value or the most common value required?
2. The frequency distribution of the data also affects the choice to be made. Selection of a location measure depends on whether the data are skewed or symmetrical. Compare the two distributions shown in Figure 3.2. Note that both

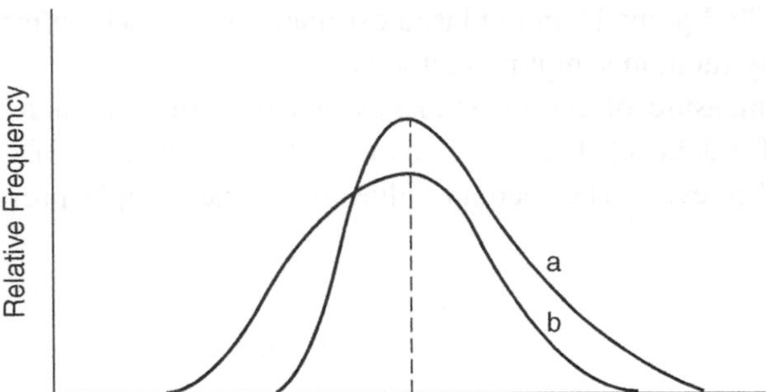

Figure 3.2 Skewed distributions with same mode, different means.

distributions have approximately the same mode but not the same mean. A normal distribution has a mean, mode, and median of equal values, while nonsymmetrical distributions differ in values for these measures.

Measures of central tendency provide only a partial description of data. Information on variation or dispersion of the data is also important for fuller description of data. To illustrate the importance of data variation, note that the following sets of numbers have the same mean but different variability, as indicated by the range in data values.

Set 1	Set 2
8, 6, 7, 8, 7, 6	3, 16, 3, 3, 8, 9
$\frac{\sum X}{n} = \frac{42}{6} = 7.0$	$\frac{\sum X}{n} = \frac{42}{6} = 7.0$

Data set 1 ranges from 6 to 8, while set 2 values range from 3 to 16. Therefore, means alone are not equally good descriptors of their respective data sets. The data of set 2 is said to have more variation than set 1. Some measure of variability or dispersion of the data would add to the information that is conveyed by a value for the mean. We may conclude, then, that a measure of dispersion is also needed to determine the reliability of an estimate of the population mean.

3.1.2 Measures of dispersion

Measures of dispersion can be obtained from the range in data values, the quartile deviation, the mean deviation, and the standard deviation. These measures may be made from all population values and, like the mean, are called "parameters." They may be made also from sample data values and are referred to as "statistics."

The range is the simplest calculation used to estimate a measure of dispersion and is the difference between the smallest and the largest values of a variable. In the sample data (for sample size of $n \leq 10$) the range r is

$$r = 50 - 22 = 28 \tag{3.6}$$

To estimate dispersion or the standard deviation s, the range is divided by 4:

$$s = \frac{r}{4} = \frac{28}{4} = 7 \tag{3.7}$$

For the population example, the range R is

$$R = 71 - 12 = 59 \tag{3.8}$$

The standard deviation σ for the population is estimated to be

$$\sigma = \frac{R}{4} = \frac{59}{4} = 14.8$$

The range may be an adequate measure of dispersion, provided the extreme values are not a great distance from the remaining values of the data group. The quartile deviation Q is defined as

$$Q = \frac{(Q_3 - Q_1)}{2} \tag{3.9}$$

where Q_1 is the value of the variable that separates the second quarter (25%) of the data from the first 25% and Q_3 separates the second 25% of the data from the third 25%. For the sample data

$$Q = \frac{46 - 29}{2} = 8.5 \tag{3.10}$$

For the population

$$Q = \frac{48 - 35}{2} = 6.5 \tag{3.11}$$

Because the quartile deviation is one-half of the range of Q_3 and Q_1 values, it is useful as an estimate of dispersion instead of R (or r) when the extremes of the data are widely separated from remaining values. The reliability of this dispersion measure (Q) depends on the concentration of values at the quartiles of the population from which the sample is selected. The presence of discontinuities at quartiles renders the measure unreliable for estimating the standard deviation.

The mean deviation is another measure of dispersion and is the mean of the absolute deviations of all variables from the arithmetic mean. A deviation is obtained by subtracting an individual value of X_i from the mean $\bar{X}$ and using the absolute value of the result in further calculations. Use of the absolute value is necessary since the sum of deviations around a sample mean will always equal zero. Let an individual deviation be represented by a lower case x; then

$$x = X_i - \bar{X} \tag{3.12}$$

where X_i is the ith value and $\bar{X}$ is the sample mean. The mean deviation md is obtained by

$$md = \frac{\sum |x|}{n} \tag{3.13}$$

As before, the Σ indicates that the summation is taken over all values in the sample, $|x|$ is the absolute value of the x in Equation (3.13), and n represents the number of observations or measurements in the sample. For the sample

$$\begin{array}{llll}
X_1 = 22 & -\ 39.3 = & -17.30 \\
X_2 = 27 & -\ 39.3 = & -12.30 \\
\cdot & \cdot & \cdot \\
\cdot & \cdot & \cdot \\
\cdot & \cdot & \cdot \\
X_9 = 48 & -\ 39.3 = & 8.70 \\
X_{10} = 50 & -\ 39.3 = & 10.70
\end{array}$$

and

$$md = \frac{80.4}{10} = 8.04 \tag{3.14}$$

For the population

$$MD = \frac{446}{50} = 8.92 \tag{3.15}$$

Note that the sample estimate of 8.04 is close to that of the population mean deviation. The mean deviation is seldom used to estimate dispersion because of the requirement to ignore the signs of differences of individual values from the mean. This problem is avoided by using the standard deviation, which is obtained as follows: (1) square the deviation of a value from the mean, (2) sum the squares, (3) divide by N, and (4) take the square root of the quotient from (3). The standard deviation for a population is represented by σ and is obtained from

$$\sigma = \sqrt{\frac{\sum_{i=1}^{N} (X_i - \mu)^2}{N}} \tag{3.16}$$

The square of the standard deviation of a population is called the "variance" and is represented by σ^2, where

$$\sigma^2 = \frac{\sum_{i=1}^{N} (X_i - \mu)^2}{N} \tag{3.17}$$

The standard deviation of a sample is represented by s and the sample variance by s^2. The usual case is to compute the standard deviation and variance of a sample in order to make an inference about the population dispersion. When a sample variance is calculated, the divisor $n - 1$ is used in place of n in order to obtain an unbiased estimate of the variance and subsequently the standard deviation.

The equation for the sample standard deviation is

$$s = \sqrt{\frac{\sum_{i=1}^{n} (X_i - \bar{X})^2}{n - 1}} \tag{3.18}$$

and the variance is

$$s^2 = \frac{\sum_{i=1}^{n} (X_i - \bar{X})^2}{n - 1} \tag{3.19}$$

In this case, s is an unbiased estimate of the population standard deviation, σ and s^2 is an unbiased estimate of the population variance, σ^2.

These equations are conceptual in nature and define the standard deviation in a meaningful way. Deviations from the mean

$$x_i = X_i - \bar{X} \tag{3.20}$$

also may be presented as standard deviation units. For example, the measurement in row 1, column 1 of Table 3.1 is 28. It deviates from the mean (41.2) by –13.2 g.

This deviation can be expressed in standard units by dividing –13.2 g by σ, which is 11.8 g. That is

$$\frac{x}{\sigma} = \text{standard deviation units} \tag{3.21}$$

In this case, we have –13.2 g/11.8 g or –1.2 standard deviation units, and the unit of measurement (g/m^2) is eliminated. Therefore, any measurement can be compared to any other measurement on this basis. For example, a measure of grams per square meter can be compared to plant cover in percentage units since standard deviation units eliminate units of measure. It can be verified that all the values for Table 3.1 lie between ± 3 standard deviation units of the mean.

In summary, measures of dispersion provide a basis for comparison of sets of data. The question regarding which measure of dispersion to use for a problem is answered in much the same way as the question regarding which average to use. The selection depends on the nature of the problem, the characteristics of the measure of dispersion, and the type of analysis required.

3.2 Principles of data behavior

Basic principles of vegetation data behavior are discussed to develop an intuitive understanding of variations encountered in the real world. However, variations that are prevalent in measurement data from vegetation systems rarely can be determined exactly. Rather, such variation must be estimated from sample data. Since a vegetation variable does not yield precisely the same value in every sample, it would appear that some degree of certainty needs to be placed on a specified value obtained from a sample process. That is, uncertainty is present in conclusions drawn from sample data collected for a given vegetation characteristic, such as biomass (g/m^2). The presence of uncertainty suggests that relevant factors are not known for sure, but there is still interest in a need to place some level of confidence on a conclusive statement. The variance estimate allows an estimate of confidence to be made as presented below.

3.2.1 Patterns of data commonly observed

In the previous discussion of variations, it was pointed out that data follow certain patterns, and an illustration was graphically presented of the pattern of the data in Figure 3.1. As can be observed, the pattern is symmetrical (bell-shaped) about a certain point. In this case, the mean value of 41.2 g/m^2 occurs at the central part of the curve. Furthermore, a conversion of the data to probabilities does not change the shape of the curve. It is still symmetrical and bell-shaped. This particular type of curve will be closely approximated quite often in nature if data are classified according to frequency of occurrence. Thus, measurements such as plant heights, biomass weights, plant moisture content, and many other variables are approximately "normally distributed." We expect the concentration of values to be at, or near, the central value (location) and to observe an approximately equal distribution (dispersion) on both sides of this central value.

Occasionally other patterns are encountered that do not follow this bell-shaped curve, i.e., they are not normally distributed. Such data occur especially in presence–absence type of data represented by a numerical value of 1 or 0, respectively, for a plant characteristic. An example is the frequency measure for a species (Chapter 5) and the data can be described by a "point binomial" distribution. These data can have only one of two possible values (0 or 1) and should be analyzed by binomial methods to estimate parameters (mean and variance).

Another type of data commonly encountered occurs when certain values are clustered together in the population, forming "islands" in the field. This kind of behavior in data is referred to as being "contagious" or "aggregated" and forms yet another kind of pattern that differs from the two previously mentioned. Numbers of individual plants in an area, for example, may be clustered under certain conditions with the result that many quadrats may contain few or no individual plants of a species while a few quadrats will contain large numbers of the species. The number of individuals of a species will depend on an environmental factor or set of factors. This kind of data may be a "count binomial," "Poisson," or one of the other aggregated or contagious distributions presented in Section 3.4. These data should be analyzed by techniques that are designed for the given distribution.

3.2.2 The normal distribution

The normal distribution and its parameters are used almost exclusively in applied statistics. The reasons for the wide use of the normal distribution in data analyses is summarized by stating that: (1) many populations (weights, heights, etc.) that occur in nature appear to be approximately normally distributed and (2) sampling and subsequent analyses of these populations are easily handled with equations used for assumptions of a normal distribution.

The frequency with which a particular value is expected to occur, if the variable is normally distributed, is calculated by the relation

$$Y = \frac{1}{\sqrt{2\pi}\sigma} e^{-(X-\mu)^2/2\sigma^2} \quad \text{for} \quad -\infty < X < +\infty \tag{3.22}$$

where Y is the frequency of occurrence and X is the variable of interest. This distribution has two unknown parameters. The parameter μ, is known as the "mean"

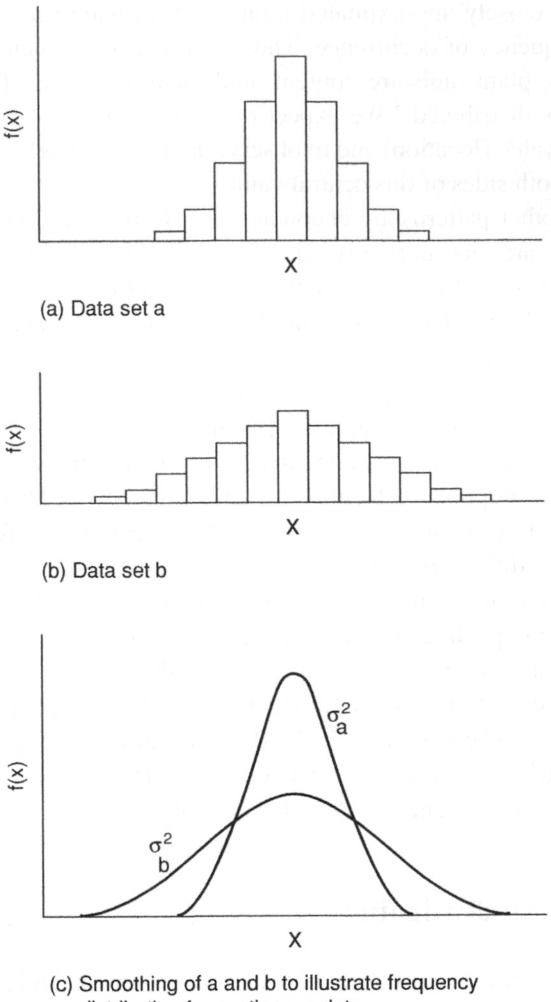

Figure 3.3 Measures of central tendency of data and variations. Since a distribution smaller in range than *b*, the variation in *b* is greater; *c* illustrates the principle with continuous data.

(location) of the distribution, while σ^2 is called the "variance" (dispersion) of the distribution. Note that the mean is located under the highest point on the curve in Figure 3.3. This is one characteristic of the parameter μ, and it measures location. The variance σ^2, on the other hand, changes the shape of the curve by spreading the same area over a larger horizontal region if σ^2 is large, while a small value of σ^2 gives a curve that is horizontally narrow with a higher vertical rise (Figure 3.3).

There are many normal distributions because there are many possible values for μ and σ^2. As a matter of fact, there is a normal distribution for each possible pair of values that could be assigned to the parameters μ and σ^2. Each population is described when: (1) the form (shape) of the frequency curve and (2) the parameters found in the frequency curve in Equation (3.22) are known. Then it follows that the normal distribution of concern is completely described if the mean μ and the variance σ^2 are known. That is, the mean and the variance of a particular normal distribution give all the information necessary about this distribution. Consequently, this is a convenient property to use in data analysis. This does not suggest, however, that the mean and variance completely describe *every* population, but it is true of populations that follow the normal distribution.

3.2.3 Sample statistics and the normal distribution

In the example, the entire 50 m^2 population was measured. Thus, values of μ and σ^2 are known for that population. These parameters are rarely known because time and financial limitations allow for measurement of only a few individual items rather than all possible values. Furthermore, it is sometimes necessary to measure a few values because the process of being measured may destroy the material sampled. For example, clipping quadrats to measure above-ground biomass removes the material measured. Therefore, sample estimates are used to estimate the population mean and variance. In review, the sample mean $\bar{X}$ is an estimate of μ, one population parameter, while s^2 is a sample estimate of σ^2, the other population parameter.

The mean and variance of a sample are calculated, respectively, in Equations (3.2) and (3.19). These two functions of sample data are known as "statistics" because they do not contain any unknown parameters (i.e., μ or σ^2). Other statistics besides $\bar{X}$ and s^2 may be useful to calculate. The mean, median, and mode are all the same numerical value for the normal distribution. This identity is not true for non-normal distributions (see Figure 3.2).

Relationships expressed by the statistics $\bar{X}$ and s^2 to their population values are "point estimates"; that is, a specific point (value) is calculated as opposed to an interval estimator. In the latter, a range of values is calculated wherein the true population parameter is expected to occur. For example, the interval for the mean may be expressed by

$$L < \mu < U \tag{3.23}$$

where L represents the lower bound and U the upper bound of μ. If we do not know the values for μ and σ^2, the interval estimation technique allows us to place "confidence limits" on estimates for values of these parameters. Then a statement can be made as to the certainty (probability) that the calculated intervals will include the population mean and variance, respectively. It was pointed out previously that

Table 3.2 The possible values of $\bar{X}$ given an example population of $N = 4$ and a sample size of $n = 3$

Sample No.		Sample Mean
1	31, 34, 37	34
2	31, 34, 40	35
3	31, 37, 40	36

the normal distribution is a bell-shaped distribution in which the mean, median, and mode all have the same value. It can be proved that if a distance of σ is measured to the right of μ, and to the left of μ, then 68.27% of the area under the normal curve will have been included. Any multiple of σ can be used, but usually 1, 2, or 3 is used in practice. From a practical standpoint these results are interpreted as follows. If an item is randomly drawn from a population of values, the probability is 0.6827 (or 68.27%) that its value will be between $\mu - \sigma$ and $\mu + \sigma$. Between $\mu - 2\sigma$ and $\mu + 2\sigma$ includes 0.954 of the values, and 0.997 of the values lie between $\mu - 3\sigma$ and $\mu + 3\sigma$. Thus, confidence limits can be placed on values of interest.

In most cases, interest is in the mean value (μ) of the population since it is an average value. Therefore, some certainty is needed to ensure that the true mean μ, has been included when μ, is estimated by the sample mean $\bar{X}$. Suppose that three observations were drawn from a population and the sample mean $\bar{X}$ was calculated. Repeat the process and obtain combinations of three observations from a population of only four values: 31, 34, 37, and 40. The means of the possible samples are given in Table 3.2. Each sample gives a mean that may or may not be equal to that of other samples. Nevertheless, the probability of obtaining a given $\bar{X}$ can be calculated if all possible values of $\bar{X}$ are known. In other words, the sampling distribution of $\bar{X}$ is the probability distribution where the possible outcomes are the values that the sample statistic ($\bar{X}$) may assume. Note that possible values of $\bar{X}$ range from 34 to 37, which is a narrower range than values of individuals in a sample. Furthermore the distribution of $\bar{X}$ is centered on 35.5, which is the average of $\bar{X}$ values. Call this value $\bar{\bar{X}}$.

The value of μ, the population mean, is also 35.5

$$\frac{31 + 34 + 37 + 40}{4} = 35.5 \qquad (3.24)$$

It is no coincidence that the sampling distribution of $\bar{X}$is centered about the population mean. The distribution of $\bar{X}$ is not as variable as the individual population samples. Thus, it is seen that $\bar{\bar{X}}$ is centered about μ, and $\sigma_{\bar{x}}^2$ (variance of $\bar{X}$) is smaller than σ^2 (the population variance of X).

The mean of the sampling distribution of $\bar{X}$, $\mu_{\bar{x}}$, is always equal to the population mean, μ, when simple random sampling is used. This simple equality can be expressed as

$$\mu_{\bar{x}} = \mu \tag{3.25}$$

This relation is very useful in statistics because it shows that the sample mean, based on a simple random sample, neither overestimates nor underestimates the population mean. Therefore, the sample mean $\bar{X}$ is called an "unbiased estimator of the population mean" μ. The term "unbiased" refers only to the tendency of sampling errors to cancel out when one considers all possible sample means that can be obtained.

The variance of the population mean σ^2 is

$$\sigma_{\mu}^{2} = \frac{\sigma^2}{N} \tag{3.26}$$

The standard error of these population means (usually referred to as "standard error" rather than as the "standard deviation of the mean") is calculated as

$$\sigma_{\mu} = \frac{\sigma}{\sqrt{N}} \tag{3.27}$$

The variance of a sample mean $\bar{X}$ is

$$s_{\bar{x}}^{2} = \frac{s^2}{n} \tag{3.28}$$

The standard error of these sample means is calculated as

$$s_{\bar{x}} = \frac{s}{\sqrt{n}} \tag{3.29}$$

Then, the standard error of the sample mean $s_{\bar{x}}$ is always equal to $\left(\sqrt{n}\right)^{-1}$ times the standard deviation (s) of the sample.

One of the most important theorems in statistics, called the "central limit theorem," states that the sampling distribution of $\bar{X}$ is approximately normal if the simple random sample size is sufficiently large. If the population sampled follows a normal probability distribution, the sampling distribution of $\bar{X}$ is exactly normal for any sample size. This theorem is given here so that the reader may have a better insight as to why the mean is used almost exclusively in placing confidence limits in data analysis. Equation (3.28) shows that the sample size n affects the size in variability of $\bar{X}$, and N affects the size in the population variability of the mean (Equation 3.26).

Thus far, the mean and variance have been defined mostly in terms of population parameters. An actual use of these parameters in a more meaningful way is needed. In review, the mean of the sample, denoted by $\bar{X}$, gives some indication of the most likely value that is expected to be encountered, should the sampling of a population be repeated again under the same conditions. The variance of the sample s^2 is the average value of the squared deviations of the individual values from the sample mean. Still another useful statistic is s, the standard deviation of the sample, which is obtained from the square root of the variance.

It was pointed out earlier that the area under a continuous frequency curve represents probability. Thus, to determine a desired probability for a value from a normal distribution, a calculation of the area under the distribution curve is made. However, these laborious calculations can be avoided with the use of a table of areas that apply to the entire family of normal distributions. This table is found in statistical texts under the title "The Standardized Normal Density Distribution". To make full use of this kind of table, standardization is made for all normal distributions. This is done by transforming the deviation of a value from its mean into standard deviation units as presented earlier. Let Z be the transformed variable calculated as

$$Z = \frac{X - \mu}{\sigma} \tag{3.30}$$

As can be seen, Z is the number of standard deviations that a value X is from its mean μ.

From the data in Table 3.1, the example of 53 g is taken from row 1, column 4

$$Z = \frac{53.0 - 41.2}{11.8} = 1.0 \tag{3.31}$$

That is, the value of 53 g/m^2 found in quadrat 31 of Table 3.1 is one standard deviation more than the average population value of 41.2 g/m^2. From a table of the standardized normal variate Z, it can be seen that the probability that a plot will have between 41.2 and 53.0 g is 0.3413. In other words, 34.13% of the quadrats measured will weigh between 41.2 and 53 g/m^2. Other intervals of interest can be calculated similarly. Usually, however, interest is in values that lie within plus or minus 1, 2, or 3 standard deviations of their mean. Therefore, probabilities of 0.682, 0.954, and 0.997, respectively, are widely used in analyzing and discussing results.

3.2.4 Confidence limits

Use of the foregoing information enables the construction of a confidence interval *CI* around the mean $\bar{X}$. If, for example, a sample of given limits includes the real population mean μ, 95% of the time, the following formula would be used

$$CI = \bar{X} \pm Zs_{\bar{X}} = \bar{X} \pm 2s_{\bar{X}} \tag{3.32}$$

where *CI* is the confidence interval and $Z = 2$, which will enable 95% of possible confidence intervals to include μ. We usually take only one sample of size n and calculate only one confidence interval.

An example of a confidence interval will illustrate the calculations. Estimates of the mean and variance from the sample obtained from data in Table 3.1 will be used. If

$$\begin{aligned} \bar{X} &= 39.3 \text{ g/m}^2, & s^2 &= 98.0 \\ s_{\bar{x}} &= 3.1 \text{ g/m}^2, & Z &= 2.0 \end{aligned}$$

then for a confidence level of 95%

$$39.3 - (2)(3.1) < \mu < 39.3 + (2)(3.1) \tag{3.33}$$

or

$$33.1 \text{ g/m}^2 < \mu < 45.5 \text{ g/m}^2 \tag{3.34}$$

Recall that $\mu = 41.2$ g/m^2 for the population represented in Table 3.1. Then this particular confidence interval, based on the sample taken, does include μ, and 95% of all possible intervals calculated from samples will also include μ.

3.3 Sample size

Sample size adequacy and its determination are of concern to those involved in vegetation inventories. Sample size is the number of observations made on a measured characteristic of vegetation (e.g., cover or biomass). In the United States, reclaimed mined land was to be evaluated by comparison of vegetation on this land with the original plant composition, as measured by cover, biomass, and shrub density. Rules and regulations concerning vegetation inventories and comparisons were established by the Office of Surface Mining of the Department of the Interior. In summary these requirements included:

> The lower limit for equality is set at 90% of the ground cover and productivity of the reference area with 90% statistical confidence, or with 80% statistical confidence on shrublands, or ground cover and productivity will be at least 90% of the stands set forth in an approved technical guide.

A review of some basic formulas from statistics may clarify the use of certain equations to estimate an adequate sample size. These are presented in a sequence that leads to appropriate uses of the well-known size estimation formula for a univariate normally distributed vegetation characteristic. Keep in mind that statistics in general involve estimation of means and variances. Thompson (1992) suggested the estimation of the population variance is usually the weakness in obtaining an adequate sample size to estimate a population mean. Then the ecologist might find it worthwhile to obtain a few observations for a measure of say, cover, by obtaining a small sample size (5–6) from areas of the plant community where cover values are apparently the largest and a few observations in areas where cover values are the lowest. Combine the two data sets to estimate the mean and variance. These two sets of samples will produce a large range in data values and subsequently a large variance to be used in one of the equations for a preliminary estimate of the necessary sample size.

An example of mined land that has been restored will be used to illustrate procedures. First, formulate a null hypothesis such that, $H_0 : \mu = k\bar{x}$, where μ is the mean of the undisturbed plant community (true population value) and $\bar{X}$ is an estimate of the mean from the restored plant community.

Let

$$t = \frac{\bar{X} - \mu}{s_{\bar{x}}} \tag{3.35}$$

Then

$$t = \frac{\bar{X} - \mu}{s/\sqrt{n}} \tag{3.36}$$

The value for t is taken from a table, and its value is dependent on the number of observations used to obtain estimates of the sample average $\bar{X}$ and the sample standard deviation s. It also depends on the confidence level desired for using $\bar{X}$ as an estimate of μ, the true population average value. This population average is estimated by $\bar{X}$ of a sample and remains unknown in most vegetation studies.

Then, by rearranging and removing the square root sign, we obtain

$$n = \frac{t^2 s^2}{\left(\bar{X} - \mu\right)^2} \tag{3.37}$$

where n is the estimated number of observations needed to obtain an estimate of the difference $\bar{X} - \mu$ within a given probability. This probability value is indicated by the selection of t, which is a value from a table of the t-distribution, and the population variance σ^2 is also assumed to be adequately estimated from s^2. Since the estimate of the mean $\bar{X}$ is also obtained from a preliminary sample of an area for vegetation production using a quadrat, n is only the first approximation to an adequate sample size. That is, the t-value depends on the size of the preliminary sample; i.e., $n - 1$, which is called the "degrees of freedom" (*df*) for estimating $\bar{X}$. Then the solution for n is an iterative process that requires n observations be obtained and that Equation (3.37) be solved again for $\bar{X}$, s^2, and t until no change in n occurs.

For example, from an ecosystem restoration project, let the value $(\bar{X} - \mu)$ be small so that it is, say, no more than $\left(0.10\bar{X}\right)$. That is, the true difference of the sample mean from the population (undisturbed plant community) mean (μ) will be no less than 10% of the sample mean $\bar{X}$. It follows that

$$\left(\bar{X} - \mu\right) = 0.10\bar{X} \tag{3.38}$$

and

$$\mu = \left(\bar{X}\right)(0.90) \tag{3.39}$$

Since μ is the true population mean and it is unknown, Equation (3.37) is estimated only by substituting Equation (3.39) into Equation (3.37). Assume that a value for s^2 is available from a previous sample. Specifically, it is assumed that vegetation characteristics, such as biomass, are normally distributed with a mean (μ) and a variance (σ^2), which are respectively estimated by $\bar{X}$ and s^2. Then

$$n = \frac{t^2 s^2}{\left[\bar{X} - \bar{X}(0.90)\right]^2} \tag{3.40}$$

or

$$n = \frac{t^2 s^2}{\left(0.10\bar{X}\right)^2} \tag{3.41}$$

In general, let $k =$ the proportion or precision that the true difference of the sample mean occurs from the population mean. Then, the general form of the equation for sample adequacy is

$$n = \frac{t^2 s^2}{\left(k\bar{X}\right)^2} \tag{3.42}$$

3.3.1 When available budget is too small

Vegetation measurements are subject to budgetary constraints as are many other components of natural resource management. Yet, experience of many field plant ecologists indicate that budgets for vegetation measurements will be among the first to be reduced or eliminated when funds are limited. Therefore, the ecologist should consider the allocated budget for measurements still can be optimized in a sample design. The above approach for sample sizes show that variance of data has the most influence over the estimated sample size. Rather than allow data variation to dictate the number of observations to make, consider fixing the costs of sampling for a measure (cover, density, etc.) and estimate the sample size that can be attained with the budget available. For a simple random or systematic sample the cost of the obtaining a sample design can be calculated as

$$C = c_0 + c_1 n \tag{3.43}$$

where C is the total cost of obtaining the measurements, c_0 is a fixed cost for data processing and analysis and c_1 is the cost of measuring one sample unit, which includes travel and measurement time from one location to the next. The sample size, n is estimated from

$$n = \frac{C - c_0}{c_1} \tag{3.44}$$

Wiegert (1962) conducted a study on the size of quadrat and the variance of the estimate of biomass to illustrate the relationship between variation in data and the cost of sampling. This reference is an example of the total costs for quadrat sampling. The relative cost equation is

$$C_f = \frac{c_f + x c_v}{c_f + c_v} \tag{3.45}$$

where the fixed cost C_f involves the costs of travel between plots, weighing, etc., c_v is the cost of time used to collect data from a quadrat, and x is the number of quadrats measured. C_f was set to the cost for an individual quadrat size relative to the cost for the smallest quadrat size.

3.3.2 Sample size for non-normal data

An estimate of sample size obtained from the equations above is useful because symmetry of the data is assumed. That is, 50% of all observational values lie below the population mean and 50% have values greater than the population mean. If this assumption is not correct and the data in fact are skewed, then these equations do not yield efficient estimators of n needed for statistical adequacy. It also might be noted that even normal approximations to an adequate sample size may not be sufficient. Then other methods might be in order for data that follow a non-normal distribution.

When data values do not follow the well-known bell-shaped curve or the preliminary data values are very small, another alternative to data transformation may be considered to estimate the sample size needed for attaining specified levels of confidence and precision. Examples that follow are data distributions that provide an adequate sample size as certain criterion is met. In particular, these distributions have a sample size that was needed to estimate a probability that an event has occurred. For example, n quadrats were observed before the first occurrence (or the rth) of a species within the quadrat. The probability of this occurrence happening can be calculated from the applicable distribution.

3.4 Data distributions

Several data frequency distributions are available to describe the various measures of vegetation characteristics. Most statistical analyses employ the normal distribution or one of the most commonly used distributions to describe vegetation data. These include the beta, beta-binomial, binomial, and Poisson. If data do not fit or are not close to the theoretical distribution known to describe such data, then the data values are not occurring at random. That is, an assignable cause is present, such as soil differences in nutrient availability, which cause plant measures to be nonrandom. For example, probability of specific density counts of individual plants occurring within a plot of given area, say, 1 m^2, can be described by one of the distributions listed above. If this is not true for a given data set, then some environmental characteristic is probably influencing the numbers of individuals that occur within a 1 m^2 area. An exploration for the cause should then be made so that the data may be fully explained.

If a data frequency distribution has a single peak, then the measurement reveals a response to a single cause. On the other hand, a bimodal distribution (two humps) is the result of the measurement responding to two or more strata such as environmental differences from one area to another. The data, then, should be categorized into the two groups before analysis is conducted and interpretation of the measurement is attempted. If the frequency distribution is relatively flat, with no real peak, the distribution is called a "rectangular distribution." The measurement then indicates that the vegetation characteristic is a response to a mixture of environmental influences and/or management practices. If data cannot be interpreted from the univariate measurement (a single characteristic, such as cover of a single species), then multivariate measurements and analysis is in order.

In many plant ecological studies preliminary data often consists of 0, 1 data values. A brief introduction is given to a distribution that describes this kind of data and then several data distributions related to this distribution are presented.

3.4.1 Bernoulli data distribution

The Bernoulli distribution describes qualitative data when only two possible outcomes can occur. For example, the use of a pin to estimate cover of a species provides for either a contact (hit $= 1$) or no contact (no hit $= 0$) on a single trial of lowering the pin to ground surface. Let x be the result of this trial; $x = 1$ for a contact or $x = 0$ for no contact with the species. Similarly, in frequency sampling, a quadrat either has or does not have an individual of a species within its boundary. As is the case for pins, these two possible outcomes can be assigned a value of zero (0) for one outcome and a one (1) for the other outcome. Determination of which outcome receives which value is arbitrary and does not affect analysis. For convenience, let the positive outcome (hit or plant occurs within the plot) be one (1) and the absence of a hit or plant from a plot be zero (0) in value. Then, for a single pin or quadrat, the Bernoulli probability distribution has only two possible values for the variable x. Therefore, a discrete distribution results and the probability that x has a given outcome is

$$P(x) = p^x q^{1-x} \quad (x = 0, 1) \tag{3.46}$$

where p is the probability that $x = 1$ and q is the probability that $x = 0$ and $q = 1 - p$. Then

$$P(0) = p^0 q^{1-0} = q \tag{3.47}$$
$$P(1) = p^1 q^{1-1} = p \tag{3.48}$$

Then the Bernoulli distribution has a single parameter p that is the probability of a success in making a hit on a plant by a pin, or finding an individual plant of interest in a quadrat. Repeated use of a number of pins (n) or examinations (n) of a quadrat will yield data to estimate p, the mean of the distribution as

$$\mu = p \tag{3.49}$$

and the variance is

$$\sigma^2 = pq \tag{3.50}$$

and the standard deviation is obviously

$$\sigma = \sqrt{pq} \tag{3.51}$$

Since $p + q = 1.0$, the variance (Equation 3.50) is maximum when $p = 0.5$ and the distribution is symmetrical. That is, when the probability of making contact with vegetation from a pin is 0.50, the variance is 0.25. As this probability value increases or decreases, the variance is always reduced in value.

If n pins are used to count the hits on vegetation, then the mean is estimated from

$$\begin{aligned}\bar{p} &= \frac{x}{n}\\ &= \frac{x_1 + x_2 + \cdots + x_n}{n}\\ &= \frac{\sum x_i}{n} \quad (x_i = 0 \text{ or } 1)\end{aligned} \tag{3.52}$$

for n pins. The variance of $\bar{p}$ is

$$V(\bar{p}) = \frac{pq}{n} \tag{3.53}$$

In almost all field sampling of vegetation with pins or quadrats an infinite population exists from a practical viewpoint so that $\bar{p}$ from Equation (3.52) provides an unbiased estimate of p. That is, a single pin provides an x value of zero (0) or one (1), while more than one pin provides for several opportunities (n, to be exact) and $\Sigma_{i=1}^{n} x_i =$ hits. Then, the probability of a single hit from three pins is (from Equation 3.46)

$$P(1) = p^1 q^{3-1} = pq^2$$

If the probability p of a hit on a single pin is 20% ($p = 0.20$), then

$$P(1) = (0.2)(0.8)^2 = 0.13$$

for any one of the three pins to contact vegetation. Because any one pin could be the one making contact, there are three ways such an arrangement could occur:

$$(0,0,1), (0,1,0), (1,0,0)$$

Then

$$= (3)(0.13) = 0.39$$

That is, the probability is 0.39 that a hit ($x = 1$) will be made by one pin if three pins are randomly placed.

To ease such computations, let the combinatorial formula be defined as

$$\binom{n}{x} = \frac{n!}{x!(n-x)!} \tag{3.54}$$

where ! is the factorial symbol, which means that $n! = n(n-1)(n-2)\ldots(1)$, define $0! = 1$, and $x = 0, 1, 2, \ldots, n$. Then for the last example

$$\binom{3}{2} = \frac{3 \cdot 2 \cdot 1}{2!(3-2)!} = \frac{6}{2} = 3$$

Equation (3.46) becomes modified to account for n observations made, each of which can take on 0 or 1. Then

$$P(x) = \binom{n}{x} p^x q^{n-x} \tag{3.55}$$

This probability function is called the "binomial probability distribution" and is discussed further below. Other sequences of Bernoulli trials (0 or 1) lead to either to a geometric, Pascal, or negative binomial distribution. This knowledge helps to explain why these distributions are often referred to in data analyses of vegetation measurements. The relationships are:

- Geometric variate: n = number of observations up to and including the first success.
- Pascal variate: n = number of observations up to and including the xth success.
- Negative binomial variate: x = number of failures before the xth success.
- Binomial variate: x = number of successes in n observations.

All observations (trials in some literature) are based on the Bernoulli parameter p and $x = 0$ or 1 at each observation point, whether the point is a pin, a quadrat, a line, or another sampling unit. Furthermore, the probability of an n or x value can be calculated and evaluated as to whether a value is significant or not.

3.4.2 Geometric data distribution

This distribution derives from the Bernoulli distribution where $(n - 1)$ is the number of pins or plots observed before a hit is made or an occurrence of a plant in a plot is noted; and the hit or occurrence gives n observations. The probability of this value of n can be calculated and adequacy is determined by the plant ecologist. The probability distribution is expressed as

$$P(n) = pq^{n-1} \tag{3.56}$$

If five pins are observed before a hit is recorded and if $p = 0.20$ from the Bernoulli as above

$$P(5) = (0.20)(0.80)^4 = 0.082$$

The probability is 8.2% that five pins are needed before the first hit is made if the Bernoulli parameter $p = 0.20$; that is, there is a 20% chance that an individual pin will contact vegetation. For application to frequency from plots, replace number of "hits" with number of plots containing the species. The mean of the distribution is

$$\bar{n} = \frac{1}{p} \tag{3.57}$$

where $\bar{n}$ is the average number of observations needed to obtain the first hit or occupied plot. The variance is given by

$$v(n) = \frac{q}{p^2} \tag{3.58}$$

3.4.3 Pascal data distribution

This data distribution can be used for sample adequacy (n) to obtain number of plots observed or pin contacts made. The probability distribution for n values is given by

$$P(n) = \binom{n-1}{n-x} p^x q^{n-x} \tag{3.59}$$

Where n is the number of observations up to and including the xth success. For instance, if interest lies in calculation of the probability that five plots are observed and two have the plant species within (xth success = the second hit), then

$$P(5) = \binom{4}{3} (0.20)^2 (0.80)^3 = \frac{4!}{3!\,1!}(0.10) = 0.08$$

or the probability is 8% that at least two plots will contain the species and the second occurrence is in the fifth plot, given the Bernoulli parameter $p = 0.20$. In other words, sample until two plots are observed to contain the species and note n (the total number of plots observed) to obtain the two occurrences. For cover values obtained from points, the number of contacts or "hits" will follow the same analysis.

The mean number of observations $\bar{n}$ (pins or quadrats) needed before the xth success is observed is given by

$$\bar{n} = \frac{x}{p} \tag{3.60}$$

and the variance is

$$V(n) = \frac{xq}{p^2} \tag{3.61}$$

3.4.4 Negative binomial data distribution

The negative binomial probability distribution results when interest in knowing the probability of observing n sample units (pins or plots), wherein the measurement is equal to zero (0) before the xth success occurs. In other words, how many absences or no-hits occur before the xth success? The answer is given by

$$P(x) = \binom{x+y-1}{y} p^x q^y \tag{3.62}$$

where $P(x)$ is the probability that exactly $x + y$ observations must be made to produce x successes, y represents the number of failures to make a hit or find a species in a plot, x is the number of successes in making a contact (hit) or finding the plant, and p and q are as defined previously. Then, if two successes ($x = 2$) and three failures ($y = 3$) occurred

$$P(3) = \frac{4!}{3!} (0.20)^2 (0.80)^3 = \frac{4}{1}(0.04)(0.51) = 0.08$$

The mean is

$$\bar{y} = \frac{xq}{p} \tag{3.63}$$

where $\bar{y}$ is the mean number of failures before the xth success is noted.

The variance is

$$V(y) = \frac{xq}{p^2} \tag{3.64}$$

Note that the mean and variance are the same as those of the Pascal distribution. The probability is very low that three pins will miss before two hits are made when $p = 0.20$ for each pin. In the case of a quadrat being the sampling unit, the probability (0.08) is very low that three quadrats will not contain the species of interest before two quadrats have the species within them.

3.4.5 Binomial data distribution

The binomial distribution results from the Bernoulli distribution with the parameter p being used repeatedly in a sample size of n to estimate the probabilities of $x = 0, 1, 2, \ldots, n$ hits by pins, quadrats occupied by a species, and so on. Then, the binomial variate x represents the number of successes encountered in n Bernoulli trials (repeated measurements for a value of 0 or 1). The probability distribution is given in Equation (3.55).

The mean number of hits or occupied quadrats expected from n observations is given by

$$\bar{x} = np \tag{3.65}$$

where n is the number of observations and p is the Bernoulli parameter. The variance is

$$v(x) = npq \tag{3.66}$$

The Bernoulli parameter p is estimated by

$$p = \frac{x}{n} \tag{3.67}$$

where x is the number of hits on plants or number of quadrats occupied by a plant of interest. Since $x = 0$ or 1, $\Sigma_{i=1}^{n} x_i = x$, the number of hits or occupied plots.

If $np > 5$ and if $0.1 < p < 0.9$, then the normal distribution provides a good approximation to x. If $np > 25$ the approximation is sufficient regardless of the value of p. So data analysis should follow that used for normally distributed data if these conditions are met.

The binomial variate x can be approximated by the Poisson variate with mean of np when $p < 0.1$. Note the following conditions are useful as a guide to determine which distribution a data set may follow if the data are discrete and based on the Bernoulli parameter p:

- Binomial variance < mean.
- Negative binomial variance > mean.
- Poisson variance = mean.

For cover estimates obtained by points, the sample adequacy equation (Cochran 1977) is

$$n = \frac{t^2 pq}{(kp)^2} \tag{3.68}$$

where t is the value from the t-table associated with the probability (confidence level) desired, p is the estimate of cover as a proportion, $q = (1 - p)$ as before, and k is the precision level with which we wish to estimate the mean cover proportion p. Note that as p approaches 0.5, and q approaches 0.5,

$$n = \frac{t^2}{k^2} \tag{3.69}$$

Then, if $k = 10\%$ of the mean is desired and $t = 1.96$ to obtain a 95% confidence in the estimate

$$n = \frac{(1.96)^2}{(.10)^2} = 384$$

A total of 384 points are needed if the cover of vegetation is near 50%. From binomial tables with the same confidence, it is noted that 250 points will detect a mean cover of 44–56% for a true cover of 50%. For $n = 100$, a range of 40–60% is obtained for the same confidence and precision levels. The sample sizes are for number of randomly located pins.

For points located randomly along transect lines, use Equation (3.42) to estimate the number of transect lines needed from

$$n = \frac{t^2 s^2}{(k\bar{x})^2}$$

where s^2 and $\bar{x}$ are obtained from transect totals. The results from Equation (3.68) dictate the number of points per transect. The optimum number of points per transects (or frames) and the number of transects should be obtained by methods presented as a two-stage sampling process. See Chapter 5 for an example of this sampling process. The binomial distribution can be used to fit a data set containing any measure of number (discrete numbers as in density of plants, number of quadrats, etc.) or of proportion, such as cover or fraction of quadrats occupied. It is note worthy that, as n becomes larger and p smaller, the Poisson distribution is an approximation to the binomial distribution. Use t-tests to test hypotheses about p.

3.4.6 Poisson data distribution

The Poisson data distribution is called the "rare-events" distribution and is a special case of the binomial distribution. That is, the probability p of an event is very small, but there are some occurrences found in a large sample (size n). If $n \geq 50$ and $np < 5$, then the event can be considered rare. For example, the observational unit (plot) is small relative to the total area. Either a plot or a pin qualifies; the plot could be 1 m^2 or 10 m^2, as long as it is small relative to the total area to be sampled.

Plant density could qualify as a Poisson variate if the number of individuals is low relative to the possible number that could grow in the area. Then, density may be 10 or 100 per unit area, as long as the value is low relative to the maximum possible. As density increases to the maximum possible, the values of density per plot will approach the frequency distribution described by the binomial distribution. The Poisson probability distribution is given by

$$P(x) = \frac{d^x e^{-d}}{x!} \tag{3.70}$$

where $x = 0, 1, 2, \ldots, d - 1, d$ (density) = numbers/unit area, e = 2.718, and $x! = (x)(x-1)\ldots(1)$ as before. Note that d could be the number of hits or occupied quadrats, as long as np is small because p is assumed to be small ($p < 0.1$).

The mean and the variance of this distribution are equal and are estimated from density d (see Chapter 7) where

$$\text{mean} = d = \frac{\text{mean number of individuals}}{\text{unit area or time}}$$

The variance of d is

$$v(d) = \frac{d}{n} \tag{3.71}$$

The mean d is the sample mean estimated as usual by $\Sigma\, x/n$. The sample variance, however, is also d, as previously stated. Again, the Poisson variate x is the limiting form of the binomial variate x as n becomes large, p approaches zero, and np approaches d.

The Poisson variate x may be approximated by the normal variate with mean and variance equal to d if $d > 9$. If the average density d is 3 individuals/m^2, then the probability of finding at least two individuals is

$$P(2) = \frac{3^2 e^{-3}}{(2)(1)} = \frac{0.45}{2} = 0.22$$

from Equation (3.70). The probability of no-hits or having no plots containing the species when $d = 3$ is

$$P(0) = \frac{3^0 e^{-3}}{0!} = 0.05$$

Use the form for the normal distribution to estimate the necessary size of sample needed by converting the Poisson variate to the normalized form if $d < 9$. That is, let

$$y = \frac{x - d}{\sqrt{d}} \tag{3.72}$$

where y will be assumed as normally distributed. Equation (3.72) should be recognized as a standardized variable from the relation

$$x_i = \frac{x_i - d}{\sigma} \tag{3.73}$$

To obtain an estimate of sample adequacy use the mean and variance of y in the sample adequacy equation for normally distributed variates Equation (3.42).

3.5 Bibliography

Cochran W. G. 1977. *Sampling Techniques*. John Wiley & Sons, Inc.: New York.

Thompson S. K. 1992. *Sampling*. John Wiley & Sons, Inc.: New York.

Wiegert R. G. 1962. The selection of an optimum quadrat size for sampling the standing crop of grasses and forbs. *Ecology* **43**: 125–129.

4

Spatial sampling designs for measurements

Requirements for the collection and interpretation of data from a study of any collection of plants and their attributes include an in-depth academic background in the sciences of plant ecology, plant taxonomy, phytosociology, and plant physiology. In addition, an understanding of plant environmental characteristics such as edaphic and meteorological factors of the area occupied by plants of interest is needed to interpret any data obtained from field measurements of plants.

Too much emphasis has been placed on data sets themselves rather than on the structure and function interface that produces plant attributes of cover, density, and biomass. This is evidenced by the primary emphasis of published works placed on quantitative measures of vegetation in general and not on plant ecology specifically. There is a need to have qualified personnel designing and collecting plant data from field studies. The need does not suggest that more emphasis should be placed on academic qualifications in quantitative materials, but individuals should have credentials in the disciplines listed above.

The single largest source of error exists in correct identification of individual plant species (a single individual) in the field. The purpose of vegetation classification will dictate the percentage of error made in a data set. If vegetation is of interest at the plant life-form level, then classification of the plant units into such a general level for measurements will obviously contain less error in the data set. On the other hand, a detailed classification of plant units into varieties of a species demands much more knowledge of plant taxonomy. Most studies concentrate on species level for partitioning of vegetation into smaller units for data collection. Shimwell (1972) presented excellent detailed discussions on the various approaches used for classification of vegetation. In any case the classification scheme should be an indication of the use to be made of measurement data.

A sampling design is the procedure used to select a sample (Thompson and Seber 1996). Sampling is used to obtain estimates of one or more of the four measures of an individual plant or groups of plants in a community. An area occupied by a plant community or other designation should be known and stated in a project description.

Measurements for Terrestrial Vegetation, Second Edition. Charles D. Bonham.

This quantity enables a reader of project reports and publications to group the size and areal distribution of collected data. In turn, one purpose of a sampling process is to produce estimates of averages and variations (variances) for measured plant characteristics of the plant community. A sampling unit can be an individual plant, a leaf, seed stalk, or a plot, quadrat, line, or point. The unit is defined according to the measure selected. Measurements of vegetation characteristics include frequency, cover, density, and biomass, and require the use of a sampling unit. These units are plots and quadrats that have relatively large areas compared to lines and points that have small areas. All measures must have a well-defined area associated with them. In turn, each sample unit is associated with a sampling design that is implemented in the vegetation area.

Most sampling designs are not identified as having a spatial component that can affect data analysis outcomes. Yet, it is difficult to imagine a plant characteristic such as biomass not being correlated with other sampling units measured in a plant community. Furthermore, in all likelihood, any set of sampling units placed within a plant community will be correlated (auto-correlated) with its set of relative *x*, *y* coordinate locations *N*. Such a correlation, when found statistically significant, indicates that the specific environment present within the community can be identified as affecting the measure (cover, density, or biomass). This correlation is often referred to as auto-correlation or cross-correlation used in time series analyses. Auto-correlation defines the quantitative behavior of an individual variable; i.e., the sinusoidal fluctuation of an individual species cover value over time or distance. Collection of data on either additional species and/or environmental factors enables the estimation of cross-correlation values between pairs of variables from each sample. The study of a plant community can employ the use of relative coordinates of (0, 0) to form a grid over the community as a starting point in the field. An example is given under Section 4.1.

It is convenient to simply refer to such sampling designs as "spatial sampling designs." If the *x*, *y* coordinates are assumed to be unimportant to interpret results, then the spatial correlation coefficient is assumed to be zero (0). This condition essentially has been assumed throughout the history of vegetation measurements.

It should be obvious that the size of the sampling unit used and its area relative to the study area determines the degree of spatial correlation that exists among measured units (e.g., percentage ground covered/m^2). It follows, moreover, that the implemented sampling design will also determine the degree of this correlation among the units. For example, if a sampling unit of 1 m^2 is used to obtain estimates of biomass in a plant community of 10×10 m (100 m^2) in size and a sample size of $n = 20$ is used to obtain these estimates, a higher degree of spatial correlation among units is likely to exist compared to results from $n = 10$. In the first case, 20% of the total population area is sampled, whereas only 10% of the area is sampled in the latter case. Sampling units have a greater opportunity to be placed further apart in the case $n = 10$ and, thus, it is more likely that the measure will not have a significant spatial correlation.

4.1 Simple random sampling

Simple random sampling is a method of selecting n units out of the N possible from a plant community or other area of vegetation such that every one of the possible sampling units has an equal chance of being selected for measurement. A sample consists of n sampling units (quadrats, points) selected on some random basis from the population (total number of units possible in the plant community).

Sample unit selection for measurement is not easily acquired from a random process implemented in vegetation field studies. If an area of 10×10 m (100 m^2) is defined as an area of ecological interest, then this relatively small area can be gridded into 100 units of 1 m^2 each; each unit can be numbered consecutively from 1 to 100. A random numbers table is then used to select a sample size of, say, 10 by drawing 10 random numbers from the range of 1 to 100 numbered units. Location of each in the field plot would be relatively easy, but not when these quadrats are located within a 1 ha area. There are now 10 000 possible quadrat locations from which a sample size of 25 need be located. Although it is possible, but not feasible, to grid the hectare into small 1 m^2 areas, one should consider other more efficient spatial sampling designs such as cluster or systematic designs (Section 4.2).

Because other sampling designs are closely related to the simple random sampling (srs) design discussed above, it is convenient to provide estimates of the mean and variance for srs here.

$$\bar{y} = \frac{1}{n}\sum_{i=1}^{n} y_i \tag{4.1}$$

where $\bar{y}$ is the sample mean, y_i is an individual measurement value (i), n is the sample size, and the variance is.

$$s^2 = \frac{\sum_{i=1}^{n}(y_i - \bar{y})^2}{n-1} \tag{4.2}$$

where symbols remain the same as defined for the same mean, except $n - 1$ in the "degrees of freedom."

Most field sampling of plant communities should only use the finite population correction factors (fpc) when the ratio $n/N \leq 5\%$ (Cochran 1977). For instance, a sample of an individual plant community of 16 ha (40 acres) using 50 quadrats of 1 m^2 area would result in a negligible fpc of <0.5% (50/160 000 = 0.03%). Plant communities restored or reclaimed after disturbances by surface mining most often result in areas of ≤16 ha when topographic features limit the extent of reseeding a large homogenous area.

Native plant communities seldom occupy large areas and when they do, stratified sampling can be used and each stratum coupled with a simple random design. In which case, Equations 4.1 and 4.2 are used to summarize stratum data.

4.2 Cluster and systematic sampling

The definition of a design using clusters of sampling units appears to be straightforward and distinct from other sampling designs. However, Thompson (1992) points out that both cluster sampling and systematic sampling follow the same basic structure; i.e., they both make use of the concept of partitioning the study area into primary units, each of which has secondary units.

A cluster sample consists of primary units with a spatial arrangement as illustrated in Figure 4.1. Each cluster consists of eight secondary units, and primary units might occur adjacent to one another as illustrated. In contrast, a systematic

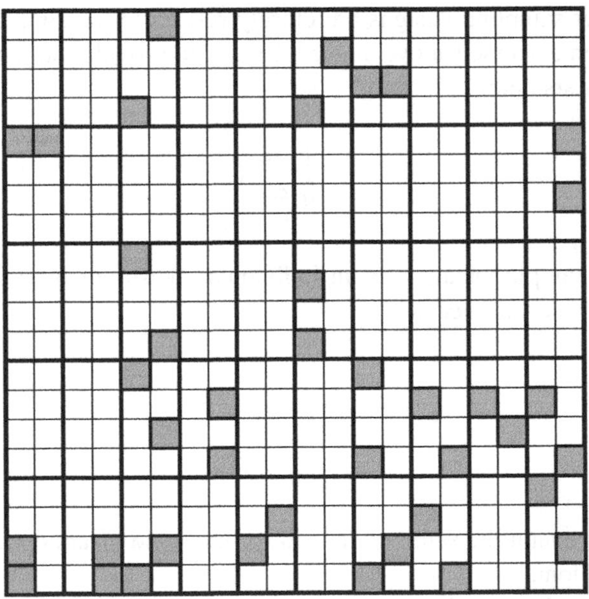

Figure 4.1 A two-stage sample of ten primary units and two secondary units per primary unit. (Reproduced, with permission, from *Sampling*. Thompson S. 1992. © John Wiley & Sons, Inc.)

sample produces a single primary unit having secondary units spaced in a systematic way across the population (plant community). Figure 4.2 shows two systematic samples as indicated by differences in their placement of the secondary units. One placement is a standard method with secondary units in a straight line. The other is a non-aligned systematic transect of secondary units placed in a random direction or distance from the systematically selected placement.

In a practical spatial way, the primary unit of a systematic sample can contain plots in a grid pattern across a plant community. Transects of points or intercepts can

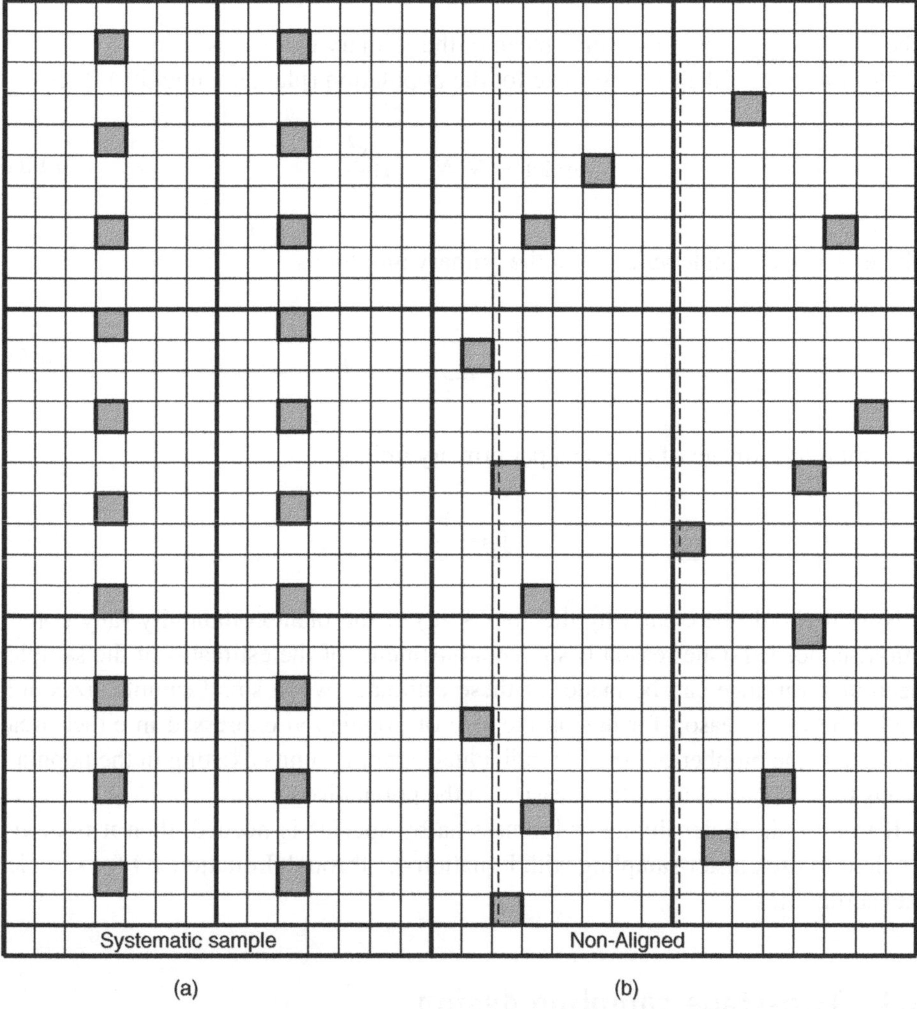

Figure 4.2 An example of two kinds of systematic sampling: (a) is a simple systematic sample, (b) is a non-aligned systematic sample. The dashed line ---- in (b) indicates where the simple design would have been placed.

also be thought of as primary units with secondary units. In which case, a transect is a primary unit that is used to analyze data collected. Recall that the transect locations are selected from the sampling design used, such as shown in Figures 4.1 and 4.2. Data from primary units only produce, for example, an estimate of the total number of plants in the study area and its variance. From Thompson (1992), these estimates are

$$Y = \frac{N}{n}\sum_{i=1}^{n} y_i = N\bar{y} \tag{4.3}$$

where $\bar{y} = \frac{1}{n}\sum_{i=1}^{n} y_i$, the sample mean of the primary unit totals.

The variance of the total estimate for the population (plant community), Y, is

$$\text{var}(Y) = N(N-n)\frac{s_u^2}{n} \tag{4.4}$$

where $\frac{s_u^2}{n}$ is the sample variance of the primary unit totals

$$s_u^2 = \frac{1}{n-1}\sum_{i=1}^{n}(y_i - \bar{y})^2 \tag{4.5}$$

An unbiased estimate of the mean per primary unit is

$$\bar{y} = \frac{Y}{N} \tag{4.6}$$

Now it should be clear why the area (size) of the plant community (any population) is needed. One reason is so that adjustments of the estimates of the sample mean and variance can be made for these estimates when small sample sizes are used. The other reason for having the size of population expressed in either area covered or the number of possible individual sampling units existing in the population is to estimate totals of the measure in the population.

If tree or shrub density/ha or biomass/ha by species is needed, then it is more efficient to use cluster sampling with Equation 4.1 through Equation 4.4 for sample data summaries.

4.3 Two-stage sampling design

Primary units are labeled as such because each of these units may be sub-sampled in one of several ways using a simple random sampling design or a systematic sampling design. That is, each primary sampling unit (psu) can be sub-divided into smaller units, and measurements of vegetation are made on each of these secondary units. For simplicity, only two sampling stages are presented (Figure 4.1). The first stage is a simple random sample of $n = 10$ from 40 possible primary units. The second stage forms a two-stage sampling design in which a single random sample of $m_i = 4$ of the $M_i = 8$ of the secondary units possible in each selected primary unit ($n = 20$).

If a plant community or other area is sub-divided into defined primary units, then secondary units might be selected from a simple random sample design and allocated within each primary unit. Other designs for sampling the secondary units could be a systematic or a stratified design. In the latter, use a post-stratification procedure for data analysis when the plant communities do not have clearly defined boundaries to form strata if they exist in the field. Secondly, outputs of analysis may show that some individual quadrats or line–point transects data more closely resemble another group (stratum) than the strata in which these sampling units were located in the field.

It should be noted that: (1) if all sampling secondary units are measured in the second stage, then the sampling design is a cluster sampling design, and (2) the simple random sampling design is often inferred from field studies to inventory and monitor vegetation types. This is clear when comparing the design to the completely random design (crd) used in experimental designs for agricultural research.

4.4 A role for satellite imagery

Satellite imagery can be used to generate maps of vegetation types and their distribution as well as for creating spatial sampling designs for vegetation measurements. Then more detailed use of the imagery can assist in locating areas for study and define boundaries for marking field plots to implement vegetation sampling. In particular, the purpose of using satellite imagery is to identify the spectral variability associated with the broad vegetation types. For instance, differences in spectral variability from one pixel to another are indicated by the various spectral bands generated by the sensors of a satellite. In turn these differences can be seen as displayed by individual pixels of approximate size 30×30 m. Plant communities in combination with topographic features can be sampled from field quadrats and related to pixels because the latter have coordinates associated with their on-ground locations. Use of these coordinates with a global positioning system (GPS) unit enables permanent location of the field sample areas for monitoring changes in vegetation and its associated abiotic variables. As a result, plant species combinations can be described with respect to environmental components as determined within pixels. For example, soil characteristics can be determined by collecting samples in conjunction with vegetation sampling units (1 m^2 for example) placed within pixels. To have a packet that depicts vegetation–environmental conditions present at the location, other environmental variables such as elevation, slope, and aspect features, as well as meteorological characteristics, should be measured within selected pixels. In turn, pixels having the same combination of variables from the set of sensors on the satellite can be grouped into "clusters" of similar vegetation–environmental responses with respect to the representation by the pixel. Then these "clusters" convey the concept of a given vegetation type (grassland, forests, etc.) consisting of plant communities that are given labels by the field ecologist.

Field sampling these communities via pixel display and locations ensures that more precise estimates are made of population parameters ($\bar{x}$, s^2) compared to a simple random sample in which samples may be allocated proportional to the size of the plant communities. Also, it may be difficult if not impossible to field map all plant communities within a vegetation type with an acceptable degree of precision. Since population parameters are weighted by the area associated with these plant communities, any error in estimation of these areas will result in biased estimates. Consideration should be given to the use of satellite imagery to locate and inventory an area of interest. Areas of pixels available from satellite imagery are precise for mapping and relocating vegetation types/plant communities. It is recognized that neither plant community nor vegetation type boundaries are specific to the area of a given pixel and that overlap of these classifications occurs with adjacent pixels. However, if adjacent pixels are classified as having the same combination of spectral bands, then any overlap difference is probably not ecologically significant.

For posterity sake, this plot (30 × 30 m) should be considered for locations of study because this spatial resolution corresponds to an existing spatial resolution of satellite imagery (a pixel representation for 30 × 30 m of ground area). As pointed out already, the location of a pixel center is available by coordinates corresponding to a GPS unit reading. Selection of psu locations can be obtained within vegetation types after random selection of pixels of similar spectral bands is made and placed on vegetation type maps of the project area. Each psu should be centered on the coordinates assigned to the psu in the imagery and placed in a north–south, east–west orientation. As stated, these large plot locations are verified by use of a GPS with an estimated accuracy of within three meters or less, depending on the satellite imagery used. More precise locations of pixel centers are available, but more costly. Locations of all psus are documented on the ground and recorded onto vegetation type maps. Future locations for newly added psus and relocations of existing field sampling units will be made more efficient by this method. Imagery from the LandSat systems is available from the 1980s and can be used to assess changes that occurred on the ground by comparison to one or more years and dates within years using the digital data.

If traditional methods are more desirable, field-generated maps of vegetation types and their plant communities can also be used for locations of the psus. Grids can be generated onto maps of the vegetation types at selected intervals with *x*, *y* coordinates. The scale of the grid should correspond to the 30×30 m so that if satellite imagery is incorporated for monitoring in the future, this imagery can be used to detect cover, density, and/or biomass changes over time with respect to a given pixel response, as indicated by spectral bands at selected locations. Some changes of plant species composition and life-form categories can be identified from satellite imagery. After grid lines are formed on maps, a non-aligned systematic grid location within a psu can be identified on the map (Figure 4.2). Select grid points with *x*, *y* coordinates to locate psus in the field. After each location is made in the field, a random direction and length of vector are chosen at the *x*, *y* coordinates. Then the field location is found and its coordinates are recorded by the GPS unit. This creates a non-aligned systematic grid and eliminates or reduces bias in analyzing data.

Note that the psu (30×30 m) corresponds to 0.25 acres on the ground and can be used for placement of transects and/or plots within the unit. Furthermore, the psu can be sub-divided into a matrix of nine 10×10 m secondary sampling units (Figure 4.3). One of these nine secondary sampling units is the standard sampling unit used to count the number of trees, measure basal diameters, dbh, and so forth in a forested plant community. One of these sub-plots (10×10 m) is also the largest plot used in plant ecological studies to form a nested plot for estimating plant cover (tree crown), shrub cover (4×4 m) and herbaceous cover (within 1×1 m) (Figure 2.9, Chapter 2). Of course, measures of densities can also be obtained from the plots for the life-forms listed for cover. Then it would be reasonable to use this nested design for estimating densities of herbs, shrubs, and trees within each of the nine secondary plots, thereby creating a cluster sampling design (all secondary units are measured) with respect to each psu plot. On the other hand, a sub-sample from one or more of the nine 10×10 m plots might be drawn for random placement of 1 m^2 plots to estimate biomass by species and thus form a two-stage sampling design for an individual psu. If pixel size (30×30 m) areas are selected at random from one spectral class (pixels) that occurs over the area of interest, then a three-stage sampling process becomes the sampling unit design (see Eberhardt and Thomas 1991, Thompson 1992).

4.4.1 Sampling units

A sampling process that is often labeled as a multistage sampling design is used to illustrate collection of data from psus and the sub-sampling thereof. The number of psus (each 30×30 m) needed to sample a large area within specified confidence levels and precision might range from 20 to 100 for an overall project area. Preliminary sampling of psus will need to be conducted and estimates of the mean and variance of measured variables obtained are to be used in the standard sample

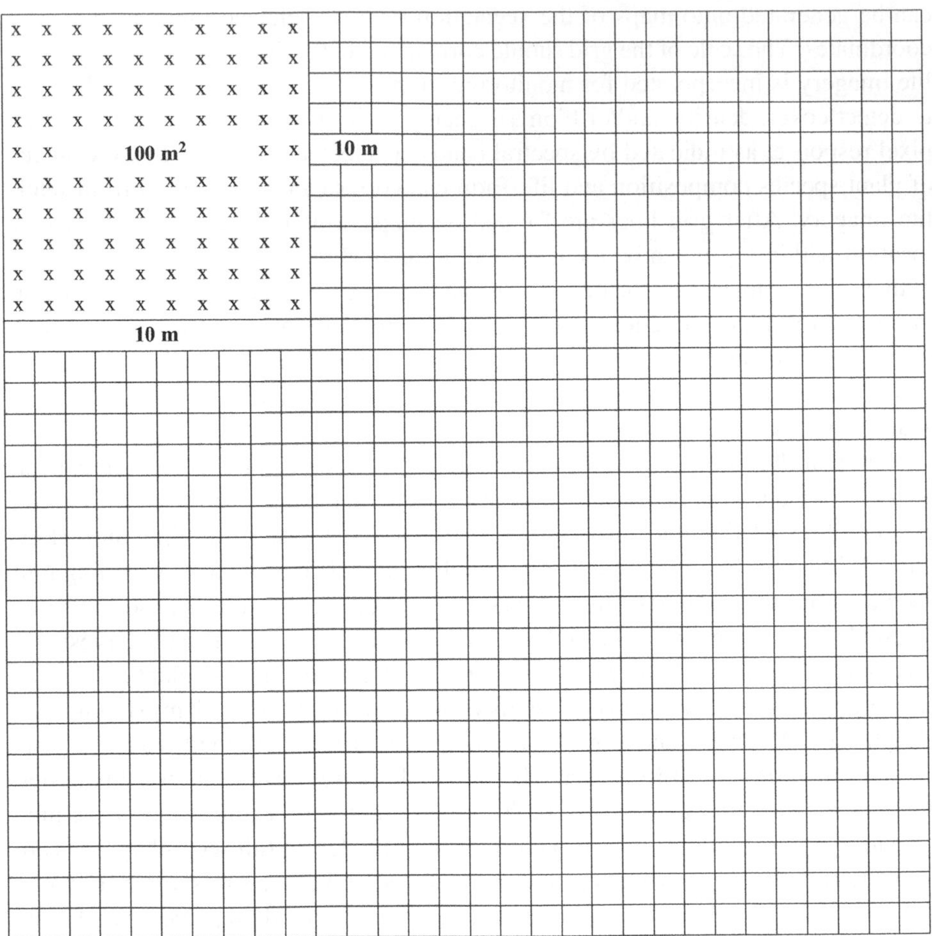

Figure 4.3 10 × 10 quadrat from 30 × 30 plot.

adequacy equation (Equation 3.42). Each psu center is the location of the pixel, by coordinates, as determined from the GPS explained earlier. This location can be marked by a metal plate attached to a metal rod, as used by some field vegetation workers to monitor vegetation changes.

Sample adequacy should be estimated within each psu if data estimates will be used for decision-making because of natural variation within this 900 m^2 area. An important consideration of sample adequacy in this field sampling design is that the design scale corresponds to Landsat satellite imagery that can be used in long-term monitoring efforts. If so, then all data collected up to the date of the current sampling will have a common field base associated with satellite imagery spatial scales.

Point–line transects measuring 10 m in length to estimate cover can be established in each of the nine secondary sampling units to estimate ground cover by species or life-forms creating a cluster sampling design (Figure 4.3). One transect is located in each of the nine secondary sampling units (10×10 m) and oriented in a north–south direction to correspond to pixel orientation. Point samples (hit/miss) should be collected every 10 cm along each of the 10 m transects for a total of 100 point estimates per transect and 900 points per psu. In general, if vegetation comes into contact with the edge of the tape at the 10 cm intervals, it will be recorded as a hit, otherwise it will be recorded as a bare soil, rock, or litter. If a random sample of five of the nine sub-units is drawn for transect placement, 500 points for the psu will be obtained and a two-stage sampling design is formed. As a side note, it is interesting to note that Thompson (1992) found that 510 points are needed to estimate cover proportions that sum to 1.00 for the sum of cover proportions in multiple categories (includes life forms). This $n = 510$ sampling units needed to estimate proportions for cover by species refers to the number of points used to obtain 0, 1 data.

Transects could be 20 m in length, but information at smaller spatial scales is lost because this length of transect transcends two secondary sampling units and the 100 points will spread over a larger area. Longer transects tend to mask the smaller, alternating scales of pattern in individual species by averaging down the number of contacts on the transect. In turn, estimates of plant cover by species can be biased because of spatial scale effects, while total cover estimates remain unbiased. Use of the extended plot from the frame described in Chapter 6 to measure ground cover by plant parts and other materials requires some type of random placements of the plots. At least 25 to 30 plots per psu are randomly placed within the psu. Sample adequacy for cover/psu can be estimated from the data obtained (Equation 3.42).

Any vegetation, soil, or other measures can be added to any existing psu (or newly established ones) over time without a problem as long as the nine sub-plots are either resampled or the same sub-plots that were selected randomly at the beginning are resampled. There is no reason to mark the location of transect lines or plots in the original inventory or in subsequent resampling. Vegetation biomass data can be obtained according to quadrat size, as discussed in Chapter 8. These plots can be randomly located within each of the nine sub-plots or a selection of a sub-set from the nine.

4.4.2 Spatial sampling

The close relationship between data collection and data analysis often causes confusion to the field ecologist namely because any presentation on data collection designs seems to be incomplete without some explanation, with equations, as to how data analysis should be conducted. On the other hand it is appropriate to present only the field design used to obtain a data set. This becomes more difficult when the data collection method includes both field collection and the use of remotely sensed data from satellite imagery. Often literature searches do not lead to detailed explanations as to exactly how either data set was obtained although the article states that both sources are integrated for analyses. Special difficulties will be encountered in an effort to repeat the study when both on-ground and satellite imagery are used to predict a biological variable of interest; namely ground cover and biomass studies of a plant community and/or a vegetation type more often than not are centered around modeling of a remotely sensed data set to use for prediction of true ground cover or true biomass. It becomes prudent to merge the imagery with an actual on-the-ground data set. For this reason and others, examples are provided below of very large areas being studied to develop methods for both data collection on the ground linked to the corresponding imagery. One way to do this is to use spectral imagery data displayed as pixels that can be associated with a geographical area of ground surface.

4.4.3 Studies of large areas

Example from Mexico

Satellite imagery from LandSat-5 ETM+ was used by Reich *et al.* (2008) to assess the usefulness of the imagery to obtain measures of tree basal area (m^2/ha), biomass (tonnes/ha), canopy closure (%), and volume (m^3/ha) among other variables in forested communities. An objective was to obtain regional and local estimates of forest vegetation variables from Landsat ETM+ imagery. A two-way nested stratified sampling design was used to randomly allocate, throughout the state, on-ground sample plots the size of satellite pixels (30 × 30 m). Spectral classes were generated from the imagery and placed within two strata defined as forested and non-forested areas. Spectral information and topographic data of the study area were obtained from satellite imageries and the Digital Elevation Model (DEM). The image was normalized (Hall *et al.* 1991) to account for differences among scenes and used for model development. Nine spectral bands 1 through 5, 6L, 6H (USGS-EROS Data Center), and 7 and 8. Spectral bands 6L and 6H were thermal bands (57 m resolution) while band 8 was a panchromatic image (15 m resolution). These latter three bands were resampled to a 30 m spatial resolution using nearest-neighbor techniques (Muukkonena and Heiskanenb 2005).

The DEM was obtained from the National Elevation Dataset (NED) as a seamless ArcInfo (ESRI 1995) grid at a 90 m resolution (Gesch *et al.* 2002). The DEM was resampled to a 30 m spatial resolution using bilinear techniques (Edenius *et al.* 2003), producing a more continuous surface reflecting gradual changes in elevation at a 30 m spatial resolution.

The primary sampling unit (psu) for ground study was a plot of 30 × 30 m that corresponded to the spatial resolution of satellite imagery (30 × 30 m, pixel size). Each primary sampling unit was partitioned into nine 10 × 10 m equal area secondary sampling units (ssu). Each psu was centered on the coordinates assigned to the pixel (i.e., the center of a Landsat EMT+ pixel). Topographic maps, the satellite imagery and GPS were used to locate field plots to a precision of 3 m.

Previous studies found that forest variables, such as forest age, forest type, basal area, tree height, density, volume, and biomass were significantly correlated to Landsat TM and ETM+ imagery (Brockhaus and Khorram 1992; Lee And Nakane 1996; Steininger 2000; Laurent *et al.* 2005). These studies have shown that Landsat TM imagery provides better predictions of some forest variables (e.g., basal area) than does a radar satellite system (Hyyppa *et al.* 2000; Lefsky *et al.* 2001) or other optical sensors of similar spatial resolution (Brockhaus and Khorram 1992).

Models using the spectral variability along with topographic and climatic variables accounted for 60 to 85% of the estimated variability in these variables on ground plots. Data processing can be conducted using the two- and/or three-stage sample designs (Cochran 1977 and Thompson 1992).

FAO example

It is expedient here to summarize procedures developed for world-wide projects involved in food production and agriculture development, plus inventory and monitoring resources that sustain food production. Therefore, a short description is here provided, as issued by the Food and Agriculture Organization (FAO) of the United Nations (UN).

The Food and Agriculture Organization of the United Nations (Ponce-Hernandez *et al.* 2004) published procedures for biomass measurements in the various vegetation layers, from ground level through to tree level. Their definition of biomass is "biomass (is) all inclusive: the total amount of live and inert organic matter above and below ground expressed in tons of dry matter per unit area." Only above-ground biomass is considered here with an emphasis on remote sensing as a measurement method. According to Ponce-Hernandez *et al.*, remote-sensing images can be used in the estimation of above ground biomass in at least three ways:

To classify ground cover into vegetation types and display a map for a given surface area. In this way, spatial variability of vegetation can be observed as relatively uniform vegetation classes. These classes are useful to identify groups of species for spatial interpolation and extrapolation of biomass estimates;

To estimate biomass as a quantitative relationship (i.e., regression equations) between band ratio indices (NDVI, GVI, etc.) or direct radiance values per pixel or as numerical values per pixel. These latter values are either direct measures of biomass or related directly to biomass, i.e., leaf area index (LAI);

To use the vegetation cover classes as a sampling framework for the location of ground observations and measurements.

This latter method should be considered by plant ecologists as a basis to implement satellite images for location of pixel-size sampling units (30 × 30 m) on the ground without any attempt to integrate imagery characteristics into results of field sampling within these pixel-sized, ground-based plots. In other words, use of satellite imagery as a guideline to locate and relocate plots to obtain vegetation measures (cover, biomass, and density) for species occurring within a 900 m^2 plot (Chapter 2).

The Ponce-Hernandez *et al.* (2004) procedures, techniques and algorithms for multispectral image classification are well-documented in the remote-sensing literature. In summary, according to these authors, the field ecologists can obtain:

Multispectral images (usually Landsat Thematic Mapper (TM) bands 1, 2, 3, 4, 5, and 7) and image enhancement (stretching and filtering), corrections (geometric and radiometric), and georeferencing (registration);

Results of created false colour composite (FCC) images from TM3 (red), TM4 (short-wave infrared), and TM5 (near infrared), for example:

1. A method for selection and sampling of "training sites" on the image, inspection of "clustering" of pixels of such training sites, and selection of a classificatory algorithm for the supervised classification (from known ground-based data);
2. A vegetation classification consisting of reflectance values derived from "training sites" and vegetation types assigned to them. Then the classification results can be applied through the classificatory algorithms to the remainder FCC image. An additional step can be taken to create a polygon of the converted image and give each polygon a name for the vegetation types sampled;
3. A map of vegetation classes related to the training sites on the ground where a vegetation cover type has been observed, recorded, and validated;
4. As a biomass estimation from measurements of standing vegetation (stem and canopy) of various strata of trees and shrubs, as well as debris, deadwood, saplings, and samples of herbs and litter fall;
5. Biodiversity assessment: plant species identification and quantification for calculation of plant diversity indices;
6. A land degradation assessment made from site measurements and observations of indicators of land degradation status.

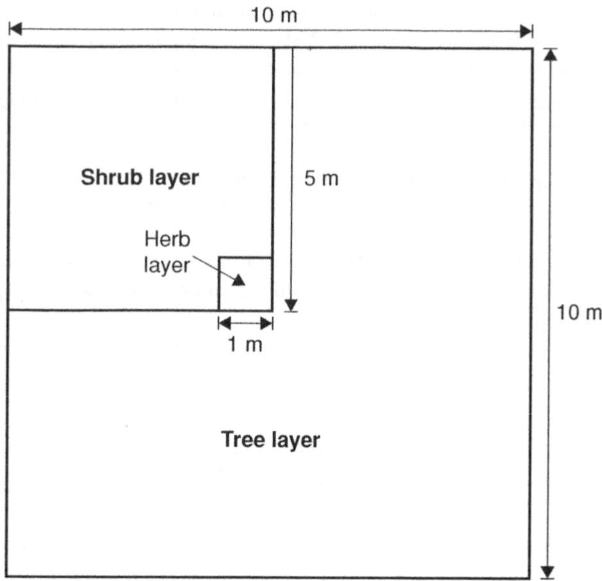

Figure 4.4 Nested quadrats recommended by FAO (2010).

The use of nested quadrats of sizes, 10×10 m, 5×5 m, and 1×1 m is recommended for measurement of biomass related to pixel imagery (Figure 4.4). The report stated that the dimensions of the quadrats selected represent a compromise between recommended ecological practice, precision, and practical considerations of time and effort as reported in the literature. Additionally, allometric and regression estimation methods used in data analysis require data as listed in Table 4.1 because these variables are considered the minimum data set for biomass estimation and they are easily obtainable and can be measured at low cost.

4.5 On-ground large-scale spatial analyses

Spatial patterns in plant species and their associated environment result in spatial heterogeneity in plant communities. The sampling design used to obtain measures will determine the degree of spatial effects detected among sampling units. Measures obtained tend to be more alike as the sampling units are located closer together than when the units are placed farther apart. A spatial auto-correlation index is one method to estimate spatial effects on the results of sampling. It follows that when biomass of a plant species is spatially correlated, then biomass of the species can be predicted at other locations in the sample area.

Systematic sampling along transect lines or on grid intersections of a study area can result in a greater auto-correlation for equally spaced quadrats. Either grid

Table 4.1 Use of each nested quadrat site for sampling and measurement. (From Ponce-Hernandez R., Parviz R., and Antoine J. 2004. *Assessing Carbon Stocks and Modeling Win–Win Scenarios of Carbon Sequestration Through Land-Use Changes*. Food and Agriculture Organization of the United Nations: Rome.) Nested quadrats are illustrated in Figure 4.4.

Quadrat dimensions	Use of quadrat in measurements and sampling
10 × 10 m	Morphometric measurements of the tree layer. Measurements of trunk and canopy of trees and large deadwood. Identification of tree species and individual organisms within a species for biodiversity assessment. Site measurements and observations for land degradation assessment.
5 × 5 m	Study of the shrub layer. Morphometric measurements of the shrub layer. Measurements of stem and canopy and small deadwood. Identification of shrub species and individual shrub organisms within species for biodiversity assessment.
1 × 1 m	Sampling of biomass of herbaceous species and grasses, above-ground and roots, litterfall and debris for drying and weighing to determine live and dead biomass. Counting of herbaceous species and number of individuals within species.

The design of nested quadrats of different sizes obeys requirements for measuring and counting vegetation of different sizes and strata, and for collecting debris and litter for estimation of biomass. The above Table 4.1 indicates the designated use for each quadrat.

points or multiple transect lines provide estimates of the index in two-dimensional space. If there is simultaneous interest in more than one variable, then a measure of the cross-correlation value I_{yx} between pair-wise variables occurring together in the sampling unit is estimated from

$$I_{yz} = \frac{\sum_{i=1}^{n}\sum_{j=1}^{n} w_{ij} y_i z_j}{W\sqrt{s_y^2\, s_z^2}} \tag{4.7}$$

where w_{ij} quantifies the degree of spatial association and can be the inverse distance between the locations i and j in the field; the value of y_i is the standardized value of variable y for quadrat i ($i = 1, 2, \ldots, n$); the value of z_j is the standardized value of variable z for quadrat j ($j = 1, 2, \ldots, n$). W is the sum of all n_2 values of w_{ij}; ${s_y}^2$ is the sample variance of y_i; and ${s_z}^2$ is the sample variance of z_j. Equation (4.7) results in a "weighted" correlation coefficient between variables y and z. Without the "weights," w_{ij}, this equation is identical to the one for r, the regular correlation measure for the two variables, y and z. Like the correlation, r, I_{yz} ranges over the interval –1 to 1.

Bonham *et al.* (1995) conducted a study of vegetation–environmental spatial variations on the Appleton–Whittle Research Ranch, Arizona, USA. The study area was defined by a mixture of shrubs, grasses, and forbs that varied across the area. Measures of vegetation-environmental variables were obtained from 100 psu (each were 10×10 m) located at intersections of a 10×10 block of intersections with 20 m spacing. Plant biomass and density were recorded for all species and life-forms present in 50 randomly located subplots, 40×40 cm, within each psu. Soil variables, pH, soil texture classes, slope, and aspect also were recorded at each psu location.

Diagonal elements of a correlation matrix formed for species measures with environmental factors like soil characteristics (pH, NO_3, etc.) are values for Moran's I statistic while the off-diagonal elements are values for the cross-correlation statistic (I_{yx}) used to describe spatial relationships of blue grama (*Bouteloua gracilis*) biomass and density with some other plant species and various site factors.

Estimates of spatial autocorrelation (I) and the bivariate cross-correlation statistic I_{ij} for selected pair-wise variables are given in Table 4.2. Estimates of the linear correlation coefficient, r, are included. Moran's I statistic on the diagonal of the matrix showed that biomass of blue grama, other grasses, and forbs were randomly distributed throughout the study area, because the spatial auto-correlations of biomass for these species were not statistically significant (NS). In contrast, the density of blue grama and hairy grama (*Bouteloua hirsute* Lag.) showed a significant positive, spatial auto-correlation with their own kind. This indicates there is some underlying spatial effect that is associated with the number of individual plants of this species over the study area. Furthermore, significant positive spatial auto-correlations were observed for soil pH, percentage clay, and elevation. Thus, one can infer that one or more factors control or influence the values of these variables over the study area.

Blue grama biomass and density, both had a significant negative cross-correlation with percentage clay content in surface soils and soil pH, and with lower soil depth pH, while blue grama density was positively correlated with percentage clay in the lower soil depth. Biomass of blue grama had a significant negative spatial cross-correlation with perennial forb biomass and a positive cross-correlation with the slope–aspect index. Blue grama's relationship with forbs can be explained partially by the changing composition of forb species as amounts of blue grama changed with respect to soil. As biomass of blue grama increased a higher number of forbs, or higher amounts of biomass of forbs also occurred.

Collection of data on plant vegetation species and their environments should follow a sampling plan that includes the relative coordinates of all sampling units so spatial auto-correlation and cross-correlation analysis can be properly analyzed. The well-established direct gradient approach used to for plant-species–environmental interpretation may be restrictive because of the single-directional approach. In contrast, vegetation responses to associated environmental factors can

Table 4.2 Spatial auto-correlation and cross-correlation analysis of species groups, and associated site characteristics. (From Bonham C. D., Reich R. M., and Leader K. K. 1995. Spatial cross correlation of *Bouteloua gracilis* with site factors. *Grassland Sci.* **41**: 196–201.)

Biomass			Density			Soils (0–15 cm)		Soils (16–30 cm)		Terrain				
Blue grama	Other grasses	Forbs	Sideoats grama	Blue grama	Hairy grama	pH	% clay	pH	% clay	Elevation	Rockiness	Slope/ aspect		
0.007[a]	−0.14	−0.024	−0.12	0.027	0.000	0.019	0.033	0.026	0.035	−0.015	−0.003	0.013	Blue grama	Biomass
NS	**	***	**	***	NS	***	***	***	***	***	NS	**		
1.0[b]	0.20	0.11	−0.08	0.35	−0.14	−0.05	−0.26	−0.00	−0.12	−0.24	−0.08	0.11		
	0.005	0.17	−0.001	−0.011	−0.011	−0.046	−0.020	−0.053	−0.020	0.059	0.009	−0.004	Other grasses	
	NS	***	NS	NS	NS	***	***	***	***	***	NS	NS		
	1.0	0.09	0.20	−0.10	−0.07	0.01	−0.11	−0.09	−0.01	0.20	0.08	−0.29		
		0.003	0.005	−0.028	−0.005	−0.043	−0.24	−0.052	−0.026	0.070	0.003	−0.011	Forbs	
		NS	NS	***	NS	***	***	***	***	***	NS	NS		
		1.0	−0.01	−0.04	−0.16	−0.26	0.24	−0.32	−0.13	0.25	−0.29	0.11		
			−0.012	−0.008	1.011	−0.022	0.008	−0.019	0.002	0.006	−0.000	−0.007	Sideoats grama	Density
			NS	**	NS	***	NS	***	NS	NS	NS	NS		
			1.0	−0.11	−0.03	0.04	−0.04	0.08	0.05	0.05	0.10	−0.21		
				0.016	0.009	0.018	0.029	0.033	0.036	−0.074	0.007	0.009	Blue grama	
				***	NS	***	***	***	***	***	NS	NS		
				1.0	−0.42	−0.07	−0.09	−0.06	0.06	−0.20	−0.31	0.09		
					0.014	−0.034	−0.018	−0.030	−0.013	−0.037	0.013	0.008	Hairy grama	
					**	***	***	***	**	***	**	NS		
					1.0	−0.09	0.10	−0.10	−0.03	−0.02	0.06	0.08		
						0.161	0.075	0.168	0.016	−0.064	−0.018	0.000	pH	Soils (0–15 cm)
						***	***	***	**	***	***	NS		
						1.0	0.37	0.90	−0.03	−0.27	−0.06	−0.13		
							0.024	0.086	0.015	−0.018	−0.012	0.002	% clay	
							***	***	***	***	NS	NS		
							1.0	0.36	0.60	−0.16	0.28	−0.15		
								0.181	0.034	−0.099	−0.023	0.004	pH	Soils (16–30 cm)
								***	***	***	***	NS		
								1.0	0.12	−0.34	−0.02	−0.16		
									0.026	−0.080	−0.003	0.008	%clay	
									***	***	NS	NS		
									1.0	−0.23	0.17	0–0.01		
										0.202	0.000	0.030	Elevation	Terrain
										***	NS	***		
										1.0	0.01	−0.16		
											−0.009	0.006	Rockiness	
											NS	NS		
											1.0	−0.23		
												−0.014	Slope/ aspect	
												NS		
												1.0		

[a]Spatial autocorrelation statistic on the diagonal and spatial cross-correlation statistic on the off-diagonal.

[b] Linear correlation coefficient, r, absolute values greater than 0.19 are significantly different from 0 at $\alpha = 0.05$.

** = significant, $\alpha = 0.01$.

*** = significant, $\alpha = 0.001$; NS = not significant, $\alpha > 0.05$.

be analyzed simultaneously over multiple directions using spatial statistical techniques.

Lake Mead burros

The influence of spatial auto-correlation on the power of an experimental design used to detect the grazing impact of wild burros on the shrub, white bursage (*Ambrosia dumosa*), was studied by Reich and Bonham (2001). Six sites dominated by the shrub were selected to represent three levels of grazing intensity (replicated) (heavy, moderate, and light) of wild burros on white bursage (Figure 4.5). At each site, sample plots ranging in size from 30 × 25 m to 60 × 25 m were permanently marked with steel t-posts. Plant canopy volume was measured because destructive measures could not be used in the study area and the *x*-*y* coordinates (m), plant

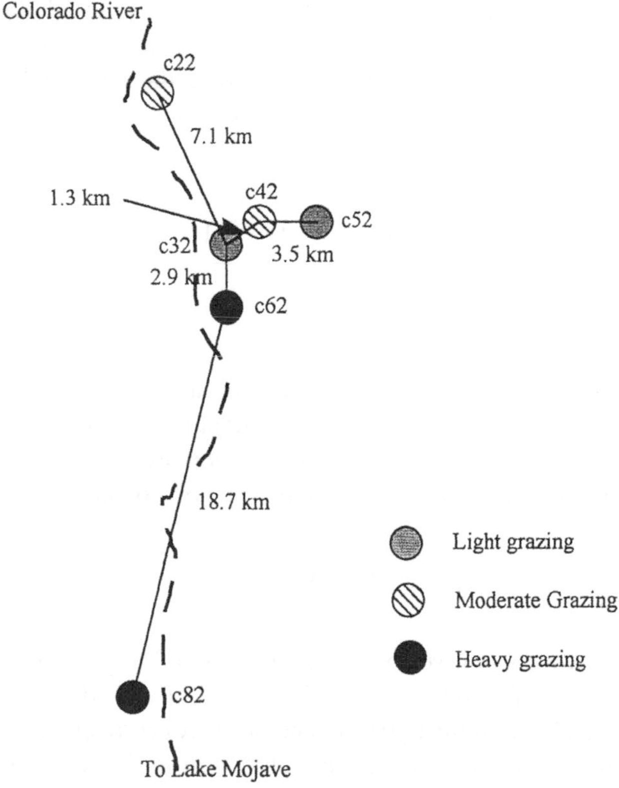

Figure 4.5 Location of sample plots of white bursage at Lake Mead Recreational Area. (Reproduced, with permission, from Spatial Analysis of Grazed White Bursage in the Lake Mead National Recreational Area, Nevada, USA. Reich, R. M. and C. D. Bonham. 2001. *Grassland Sci.* **47**(2): 128–133. © John Wiley & Sons, Inc.)

height (m), minimum crown diameter (m), and maximum crown diameter (m) were recorded for each plant on sample plots. An individual plant volume was estimated from measurements of an ellipsoid's maximum and minimum crown diameters multiplied by the height of the plant canopy above ground to yield a volumetric measure. Distances of the sample plots from the Colorado River and latitude (UTM coordinates) were obtained from a GPS unit and elevations (m) were obtained from United States Geological Survey (USGS) topographic maps of the study area.

Spatial auto-correlation analysis was used to estimate the effects of spatial pattern on canopy volumes within each of the sample plots. Moran's bivariate cross-correlation statistic I (Equation 4.7) was calculated to test the hypothesis ($p < 0.05$) that the distribution of volumes on a given sample plot was spatially independent. If the test was rejected, shrub volumes were considered to be auto-correlated; i.e., the volume sizes were affected by the location (x, y) within a plot.

An analysis of covariance model was used to test the null hypothesis of no significant differences in plant volumes due to grazing intensity. Plant volumes varied across grazing intensities that were possibly correlated with additional variables placed in the model for error reduction: D was the distance to water (river), E was the elevation, L was the change in latitude relative to the southern boundary of the study area. These variables were expressed as deviations from their overall means and the final model was given by

$$Y_{ij} = \mu + \tau_1 Z_{ijl} + \tau_2 Z_{ij21} + \beta_1 \left(D_{ij} - \bar{D}..\right) + \beta_2 \left(E_{ij} - \bar{E}..\right) \beta_3 \left(L_{ij} - \bar{L}..\right) + \varepsilon_{ij} \tag{4.8}$$

where Z_1 equals 1 if the observation was from the lightly grazed treatment, –1 if the observation was from a heavily grazed treatment, and 0 otherwise, Z_2 equals 1 if the observation was from a moderately grazed treatment, –1 if the observation was from a heavily grazed sample treatment, and 0 otherwise.

The covariance model (Equation 4.8) in matrix notation is

$$Y = Xb + e \tag{4.9}$$

where Y is an $n \times 1$ column vector of response variables $(Y_{11}, Y_{12}, \ldots Y_n)''$, b is a 6×1 column vector of regression coefficients $(\mu, t_1, t_2, \beta_1, \beta_2, \beta_3)''$, X is an $n \times 6$ matrix of zeros and ones (either positive and negative) corresponding to treatments and covariables, and e is an $n \times 1$ column vector of errors. If the errors were spatially correlated, the relationship between Y and X, was modeled (Reich and Bonham 2001) as

$$Y = Xb + (I - \rho W)^{-1} e \tag{4.10}$$

Table 4.3 Spatial autocorrelation of plant volumes (m^3) of white bursage at Lake Mead National Recreational Area by grazing intensity. (From Reich R. M. and Bonham C. D. 2001. Spatial analysis of grazed white bursage in the Lake Mead National Recreational Area, Nevada, USA. *Grassland Sci.* **47**: 128–133.)

Grazing Intensity	Plot code	Moran's I	Test Statistic	p-value (two-tail)[a]	G-section p = value[b]
Light	c32	–0.004	0.277	0.781	0.0
	c52	0.049	5.136	0.0	
Moderate	c22	–0.005	0.216	0.829	0.331
	c42	–0.028	–0.508	0.611	
Heavy	c62	–0.055	–0.897	0.370	0.840
	c82	–0.021	–0.631	0.528	

[a]Under randomization assumption.
[b]Test of the null hypothesis that the spatial autocorrelations of the distribution of volumes on plots with the same grazing intensity are not significantly different.

where W is an $n \times n$ spatial weights matrix expressing the spatial relationship among the residuals, I in matrix algebra is an $n \times n$ identity matrix, and ρ is a spatial auto-correlation coefficient indicating the level of the auto-correlation error term. An iterative procedure was used to minimize the reduced log-likelihood function that estimated ρ, because ρ varied over the interval of –1 to 1 (see Reich and Bonham 2001).

Scheffe's (1953) method was used to test for all possible comparisons among the treatments when the null hypothesis for either the classical or spatial autoregressive covariance model was rejected.

Sample plots were spread out over 28 km along the Colorado River (Figure 4.5). Results of spatial auto-correlation analysis are presented in Table 4.3. Only one plot (c52) showed a significant positive spatial auto-correlation in the distribution of individual plant volumes within this treatment; a lightly grazed replicate. Results indicated that the replicates for the lightly grazed treatments were significantly different in spatial distribution of individual plant volumes, while volume distributions within the other two treatments were not different.

4.6 Bibliography

Bonham C. D., Reich R. M., and Leader K. K. 1995. Spatial cross correlation of *Bouteloua gracilis* with site factors. *Grassland Sci.* **41**: 196–201.

Brockhaus J. A. and Khorram S. 1992. A comparison of spot and Landsat-TM data for use in conducting inventories of forest resources. *Int. J. Remote Sens.* **13**: 3035–3043.

Cochran W. G. 1977. *Sampling Techniques*, 3rd edn. John Wiley & Sons: New York.

Czaplewski R. L. and Reich R. M. 1993. Expected value and variance of Moran's bivariate spatial autocorrelation statistic under permutation. USDA Forest Service Research Paper RM-309, Fort Collins, CO.

Edenius L., Vencatasawmy C. P., Sandstrom P., and Dahlberg U. 2003. Combining satellite imagery and ancillary data to map snowbed vegetation important to Reindeer *Rangifer tarandus*. *Arct., Antarct. Alpine Res.* **35**: 150–157.

Eberhardt L. L. and Thomas J. M. 1991. Designing environmental field studies. *Ecological Monographs* **61**: 53–73.

ESRI 1995. *ARC/INFO® Software and on-line help manual*. Environmental Research Institute, Inc.: Redlands, CA.

Gesch D., Oimoen M., Greenlee S., Nelson C., Steuck M., and Tyler D. 2002. The national elevation dataset. *Photogram. Eng. Remote Sens.* **68**: 5–32.

Hall F. G., Strebel D. E., Nickerson J. E., and Gotz S. J. 1991. Radiometric reflection: Toward a common radiometric response among multidata multisensor images. *Remote Sens. Environ.* **35**: 11–27.

Hyyppa J., Hyyppa M., Inkinen M., Engdahl M., Linko S., and Zhu Y. H. 2000. Accuracy comparison of various remote sensing data sources in the retrieval of forest stand attributes. *Forest Ecol. Manag.* **128**: 109–120.

Laurent E. J., Shi H., Gatziolis D., Lebouton J. P., Walters M. B., and Lie J. 2005 Using the spatial and spectral precision of satellite imagery to predict wildlife occurrence patterns. *Remote Sens. Environ.* **97**: 249–262.

Lee N. J. and Nakane K. 1996. Forest vegetation classification and biomass estimation based on Landsat TM data in a mountain region of west Japan. In Gholz H. L., Nakane K., and Shimoda H. (eds) *The Use of Remote Sensing in the Modeling of Forest Productivity*. Kluwer: Dordrecht, The Netherlands, pp. 159–171.

Lefsky M. A., Cohen W. B., and Spies T. A. 2001. An evaluation of alternate remote sensing products for forestry inventory, monitoring and mapping of Douglas0fir forests in western Oregon. *Can. J. Forest Res.* **31**: 78–87.

Mielke P. W. Jr. and Yao Y. 1990. On g-sample empirical coverage tests: Exact and simulated null distributions on test statistics with small and moderate sample sizes. *J. Stat. Comp. Simul.* **35**: 31–39.

Muukkonena P. and Heiskanenb J. 2005. Estimating biomass for boreal forests using ASTER satellite data combined with standwise inventory data. *Remote Sens. Environ.* **99**: 434–447.

Ponce-Hernandez R., Parviz R., and Antoine J. 2004. *Assessing Carbon Stocks and Modeling Win–Win Scenarios of Carbon Sequestration Through Land-Use Changes*. Food and Agriculture Organization of the United Nations: Rome.

Reich R. M. and Bonham C. D. 2001. Spatial analysis of grazed white bursage in the Lake Mead National Recreational Area, Nevada, USA. *Grassland Sci.* **47**: 128–133.

Reich R. M., Aguirre-Bravo C., and Briseno M. A. M. 2008. An innovative approach to inventory and monitoring of natural resources in the Mexican State of Jalisco. *Environ. Monitor. Assess.* **146**: 383–396.

Scheffé H. 1953. A method for judging all contrasts in the analysis of variance. *Biometrika* **40**: 87.

Shimwell D. W. 1972. *The Description and Classification of Vegetation*. University of Seattle, Washington Press: Seattle, WA.

Steininger M. K. 2000. Satellite estimation of tropical secondary forest above-ground biomass: Data from Brazil and Bolivia. *Int. J. Remote Sens.* **21**: 1139–1157.

Thompson S. K. and Seber G. A. E. 1996. *Adaptive Sampling*. John Wiley & Sons, Inc.: New York.

Thompson S. K. 1992. *Sampling*. John Wiley & Sons, Inc.: New York.

USDA Forest Service 1988. *Range Plant Handbook*. Dover Publications, Inc.: New York.

5

Frequency

5.1 Brief history and definition

C. Raunkiaer, a Danish botanist, used the Danish noun "frekvens" and its verb form "frekvenstere" to define what he referred to as species "frequency." These words were translated to mean, respectively, "attendance" and "to frequent or attend". According to Brown (1954), Raunkiaer (1909–10, 1934) described vegetation of an area by observing the presence of all species in small sampling units of dimensions 316×316 mm (0.1 m^2) placed over the area. He suggested the importance of a species could be determined by the number of sampling units with the species present. A definition of the term "sampling units" is in Chapter 4. Such a unit could be a small plot or an entire plant community. If the sampling unit is a plant community, several communities would be examined for occurrence of plant species of interest. A complete species list would be compiled for each community and the average species frequencies would be calculated.

Frequency from its early history has been associated with an area of ground enclosed by a plot. But with the passage of time, frequency as a measure of an individual species occurrence/unit area has been replaced by point sampling without reference to contacts/unit area. As a result, current vegetation measures often do not separate a measure of frequency from that of cover. A point process does not measure frequency because units of measure are not associated with the sampling unit area; the area is assumed to have become infinitesimally small as a point. The assumption was made to eliminate effects of plot size and shape on cover data (Greig-Smith 1983). So, only cover is measured.

Therefore, only frequency measurements that use plot areas are presented in this chapter. It is important to maintain the original definition that frequency is a measure associated with an area of ground. Then, it is obvious that frequency is related to the vegetation measure of density (number of individuals/unit area) of the species. This is only true when density sampling is based on the criteria that the species is rooted within the sampling unit area and the frequency of a species is counted as presence (1) or absence (0) in that plot area. The relationship between frequency and density is presented in Chapter 7. A concise description of frequency and its

Measurements for Terrestrial Vegetation, Second Edition. Charles D. Bonham.

measure along with the other three measures is provided by the USDA (1999) for land management agencies in the USA, the University of Arizona, and the University of Idaho. Other authors have provided detailed results of various studies of frequency measurements (Greig-Smith 1983, Hironaka 1985, West 1985).

Many descriptions of vegetation using frequency measures often state that frequency indicates species abundance and distribution of a species. As pointed out in Chapter 1, the term "abundance" in plant ecology is a unit area measure of the average number of individuals in a plot when the species occurs in that plot (Greig-Smith 1983). Then frequency is related to abundance, as are the other measures and all have quantitative values per area of the sampling unit. Such an area does not refer to the areal extent of the plant community or other area. The plant ecology definition of abundance is used here.

Frequency is also thought to describe the "distribution" of a species. One assumes that the term distribution as used in the literature is in reference to the frequency value of a species, being distributed over the sample area without stating the extent of the area studied. Again, results of a measure should always be given relative to area. The definition of distribution as used for data values is given in Chapter 3. This definition is directly related to data analysis of frequency counts of presence in sampling units. Examples can be found in both plant and animal literature (Mayland 1972 and White and Bennetts 1996).

Archibald (1952) showed there is a relative effect of both plant size and plot size on estimates of frequency. Aberdeen (1958) followed up on Archibald's studies by showing mathematically how quadrat size, plant unit size, density of plants, and frequency values could be expressed in a single equation if both plants and quadrats followed a random distribution. Equation (5.1) shows this relationship.

$$a = \mathrm{e}^{-\pi\,(R+r)^2 d} \tag{5.1}$$

where a is the proportion of quadrats from which the species is missing $(1 - F)$, R is the radius of the circular quadrat, r is the radius of a circle covered by a plant unit, and d is the density of the plant units. A circular plot makes it easier to model the function describing absence of a plant since plant units are usually circular in form.

The frequency measure, when used as a relative estimate of either density or plant cover, indicates the relative importance of a species in a plant community (Aberdeen 1958, Greig-Smith 1983). Greig-Smith (1983), in particular, developed a mathematical model to describe the relationship between frequency and density when individuals of a plant species are randomly distributed within a community. In field studies, a plant species measure more often than not follows a non-random distribution because of spatial patterns at several scales within a plant community. The use of concentric plots as described and illustrated by Morrison *et al.* (1995) should be considered for frequency data or other measurement data to account for effects of spatial patterns on species attributes (Figure 5.1).

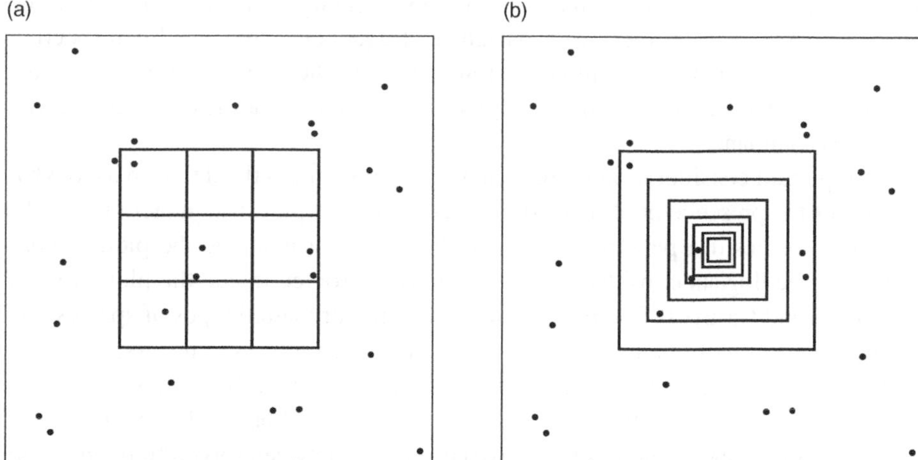

Figure 5.1 Nested quadrats for frequency. (a) A quadrat of nine contiguous equal-area (2.2 unit^2) sub-quadrats to estimate frequency. (b) A compound qyadrat of seven nested sub-quadrats to estimate frequency and importance scores. (Reproduced with permission from Morrison *et al.* 1995; An assessment of some improved techniques for estimating the abundance (frequency) of sedentary organisms. *Vegetation* **120**(2): 131–145 DOI: 10.1007/BF00034343.)

The assumption of randomness of plant frequency data allows calculation of probabilities of the number of times a species will occur in randomly placed quadrats. The probability distribution of this kind of data follows the discrete Poisson distribution that is described by a series of terms. The first term of the series, $(1 - e^{-m})$, is the proportion of quadrats that contains at least one individual, where $m =$ the average number of individuals per quadrat (unit of area). The frequency count of a species present in a quadrat is the minimum number of occupied quadrats used to estimate abundance and density of a species, as defined above.

The frequency measure is widely used by ecologists to monitor changes in vegetation because of its simplicity and speed of obtaining the data. Frequency is more efficient than any other measure for detecting change in vegetation structure (species composition) because of the time difference needed to obtain other measures. On the other hand, measurements of other characteristics are needed in addition to frequency to provide a complete analysis of the nature of the change detected.

5.2 Estimates of frequency

The term "composition of vegetation" usually implies a list of plant species that occur in a particular vegetation type. In early methods for plant community descriptions, a simple species list did not convey enough information to describe the community. So, species were assigned "frequency symbols" by inspection (e.g., rare, occasional, and common); however, soon these symbols were not sufficient to understand the inter-species dynamics. Numerical ratings were then given to species abundance.

An important consideration in frequency measurements is the decision as to what constitutes the presence or absence of a species within a plot. Raunkiaer (1909–10) counted a species if a perennating bud of the species was inside the plot. In contrast, many ecologists consider that a plant must be rooted inside the plot (Romell 1930, Cain and Castro 1959). In the case of creeping or matted types of species, the rooting criterion will result in a lower frequency. Therefore, it is necessary to distinguish between two different usages by referring to "rooted frequency" or "shoot frequency" (Greig-Smith 1983). Rooted frequency should be used to estimate frequency in plots because there is a positive correlation between rooted frequency and density for randomly distributed populations. Similarly, there is a positive correlation between shoot frequency and cover (Greig-Smith 1983), but it has been shown that the term "shoot frequency" is a misnomer applied to frequency. It refers to an estimate of cover, not frequency.

The basic measure (1, 0 for presence/absence) of species frequency from a sample of plots can be expressed as:

1. The number of plots a species occupies. This estimate of frequency is the only one having data directly obtained from field samples and is the only measurement that is a random variable by definition. The other two variables described below are based on synthetic data that result from a transformation of the direct field data. As stated in Section 5.1 the count of plots containing an individual species is a direct measure of frequency of this species, as originally defined by early plant ecologists working in Europe. The count of such sampling units is

$$F_c = \Sigma C_i \tag{5.2}$$

where F_c is the "frequency" count over n plots within a plant communities; i.e., C_i is the presence (1) or absence (0) data for the species from n plots. The sampling unit can be a plot of any size or shape and only presence data is recorded for each species.

2. The proportion of plots that has a species of interest. Frequency, as a proportion, is estimated from a random sample of plots as

$$F_p = \frac{\text{number of plots occupied}}{\text{total number of plots examined}} = F_c/n \tag{5.3}$$

where F_p is the frequency proportion that results from dividing Equation 5.1 by the total number of sampling units (n). F_p is a synthetic variable, meaning that it was created from the original data set as indicated above. This estimate is referred to as relative frequency of occurrence of a species because it is the mean frequency (expected value) and estimates the probability of occurrence (Jongman *et al.* 1995).

3. The percentage of plots occupied by a species. Multiplication of F_p from Equation 5.3 by 100, of course, will convert the frequency proportion estimate to a percentage, i.e., the percentage of sampling units containing at least one or more individuals of the species of interest. An estimate of percentage frequency is

$$F_{pc} = F_p \times 100 \tag{5.4}$$

where F_{pc} is likewise a synthetic variable created from Equation 5.3. By definition, frequency F_{pc} is the percentage of the n sampling units that contained the species. This estimate is the most commonly used frequency measure. The value of F_{pc} also has been referred as the "frequency index" or "degree of frequency" by plant ecologists (Brown 1954).

It is obvious that there are different ways of presenting the same data obtained from field sampling. In all cases, the basic data values recorded consist of 1 or 0 for presence or absence of the species in a plot. The historical definition was based on counts of presence in a plot to define frequency; these counts can be directly analyzed by use of the count binomial or other discrete data distributions, depending on the assumptions made. Otherwise, the frequency proportion data, like proportion of cover (Chapter 6), can be analyzed by the point binomial (Bonham and Clark 2005), negative binomial, beta, or the beta binomial distributions (Chen *et al.* 2007, Shiyomi *et al.* 2004). Note that the use of a proportion for frequency data analysis does not imply that the data was obtained from a point method. Methods for analysis explained by these latter authors were developed for cover estimates obtained from point data, but they also illustrate mathematically why cover and frequency estimates from points yield the same data set. However, as stated above, points do not have area and cannot be used as estimate of frequency.

5.2.1 Influences of size and shape of the plot

Species frequency is considered to be the most difficult vegetation characteristic to interpret. Greig-Smith (1983) and others thought interpretation of the data is difficult because it is not a measure independent of shape and size of the plot used to obtain data. It is even more important to realize the modern terms "frequency percentage or frequency index" are synthetic variables; they are not measured directly in those units, but rather are calculated from the field data. Results may be further confounded by the species abundance and patterns of plant growth in an otherwise homogeneous vegetation cover. If the plot unit is too small, the likelihood of a species being recorded is small, and the results of sampling indicate a low frequency value for more common species. Less common species may not be recorded at all. On the other hand, if the plot is large, results of sampling may show similar frequency for species that have different distributions in a community. Individuals of a species may, in fact, show aggregation. Therefore, results of frequency determination with a rectangular, square, or circular plot of the same area may differ. Furthermore, frequency of species determined by different sizes and shapes of the plots in two communities or in the same community at two different points in time cannot be compared because results are dependent on the shape and size of plots, as well as sample size. The same area sampled by the units has to be equivalent. Refer to Section 3.3 on Sample Size for methods to mitigate plot size and shape, as well as sample size and spatial effects on frequency data.

5.2.2 Requirements for valid comparisons

Frequency is considered a useful tool for comparison of pair-wise plant communities for similarities in occurrences of species combinations and to detect changes in vegetation structure of these plant communities over time. For example, a restored native plant community can be compared to an undisturbed plant community containing similar species. To make a valid comparison requires the sampling units to include the same sampling area in each set of plant communities that are compared. For example, the total area covered by the sampling unit has to be equal in the two areas compared. If a sample size of $50 \times 1\ m^2$ plots is used in one area and a $0.5\ m^2$ plot is used in another area, then sample size in the latter area has be 100 so that both are sampling $50\ m^2$ of the study areas. Studies of changes in frequency of species in different plant communities often have violated this requirement because species frequency was obtained from different areas sampled by plots within two or more plant communities. This occurs from the use of different plot sizes and/or different sample sizes in the two communities. A method by Helm (2010) to compare frequency data taken from different plot sizes is presented below.

5.3 Minimal area and frequency

Plant ecologists know to use the optimum size of a plot for vegetation measurement and they emphasize sampling in recurring plant assemblages that are referred to as plant communities. Historically, the optimum plot size was one that represented the species composition of the community and such a plot was referred to as having a "minimal area." It is now known that the minimal area of a plot can be determined only in a community that is relatively homogeneous in plant cover distribution. But even so, there is a requirement for a quantitative description of vegetation to be obtained that will represent the entire community. To do this usually dictates that a large number of observations with a smaller plot size be used rather than a few large minimal area plots. The large number of small-area plots ensures that the area of sampling, and, thus, vegetation conditions, have been adequately covered in the sample.

5.4 Plot size considerations

Techniques for determining optimum plot size for a measure of frequency should not be subjective. This is especially true if one is interested in finding a frequency index (frequency percentage) for all the species in a community. In this case, a plot of any size will include some species for estimates of frequency. The aim should be to sample with precision and to describe species associations within the plant community. Sample size may mitigate most perceived problems associated with frequency estimation and effects of plot size on that estimate. Consider the fact that area covered by the sampling unit, say a 0.25 m^2 quadrat, will approach an unbiased estimate of a species frequency as sample size increases, covering more of the community area. In which case, effects of spatial scales on the estimate are in proportion to scale size.

The size of sampling unit for frequency measurement is basically a function of spatial patterns of individual plants, plant size, and species richness in a sample area. Experimental data have shown that up to 10 species per sampling unit can be counted conveniently. Therefore, the size of the sample unit will vary, within limits, and will depend on the type of vegetation sampled. Cain and Castro (1959) suggested the following empirical sizes:

Moss layer	0.01–0.1 m^2
Herb layer	1–2 m^2
Tall herbs and low shrubs	4 m^2
Tall shrubs and low trees	10 m^2
Trees	100 m^2
All measures	900 m^2

The plot size of 900 m^2 is added because it relates to a satellite image with a pixel size of 30 $\times$ 30 m, as occurs at ground level, and such imagery is now used to conduct vegetation surveys. The use of one of more of the smaller plots listed above can form a sample design of sub-plots to obtain any or all of the four measures of frequency, cover, density, and biomass. Some designs used in field sampling are presented in Chapter 4.

Plot sizes listed above are found in use throughout the world, not only for frequency measures of species, but also are employed for obtaining other measures. For example, the 100 m^2 (10 × 10 m sub-plot) is a subdivision of the 30 × 30 m (satellite pixel size from LandSat imagery) as used by Reich *et al.* (2008) to survey large areas of forest types in Mexico. There are nine of these sub-plots that are possible and can be thought of as a contiguous lattice on the ground within a pixel. If all nine plots are measured for the variable of frequency, then a cluster sampling design exists. Plot sizes and sub-plots within them for the other measures are found in Chapter 3, as well as in the individual chapters for the measures. Plot sizes listed above might not provide adequate estimates of the measure of frequency. In such a case, the plant ecologist may want to consult the special cases below to determine adequate sizes to use for frequency when data is available for a given plot size.

The above listed sampling unit sizes can be used to initiate the plot arrangement for a plant community to be sampled for frequency data. Hyder *et al.* (1965) used the logarithmic relation between density and frequency to calculate the sizes of complementary plots for a nested design. Such a design would include at least two plots of different sizes; one is located within the other. Given that a random distribution for plants of a species occurs within a plant community, then density d per quadrat is given by the Poisson distribution (Greig-Smith 1983) and

$$d = -\ln\left(1 - \frac{p}{100}\right) \tag{5.5}$$

where ln is the natural logarithmic function, d is the number of plants per plot area (density), and p is the frequency percentage.

If the density of a species per unit area is known, then its frequency percentage (p) for a plot of size a is estimated as

$$p = 100\left(1 - e^{-ad}\right) \tag{5.6}$$

$$\frac{p_1}{p_2} = \frac{1 - e^{-a_1 d}}{1 - e^{-a_2 d}} \tag{5.7}$$

where p_1 is frequency associated with unit area plot a_1, p_2 is the frequency with another size of plot a_2, we should be able to determine the size of plot a_2 for the desired frequency p_2. Equation (5.7) can be written as

$$a_2 = \frac{\ln q_1 - \ln q_2 + da_1}{d} \tag{5.8}$$

where

$q_1 = 100 - \mathrm{p}_1$
$q_2 = 100 - \mathrm{p}_2$
d = density of a species

Let the area of a small plot a_1 be one unit (i.e., 1 m^2), and if the frequency of a species p_1 is 5% (the lower acceptable marginal limit), then from Equation (5.5) we have

$$d = -\ln(1 - 0.05) = 0.05 \text{ plants/m}^2 \quad (5.9)$$

Now that we know the density d for plot a_1 size 1, in which a species has a frequency of 5%, we can calculate the size of a larger plot a_2 from Equation (5.8), which gives a frequency of 95% (the upper marginal limit)

$$a_2 = \frac{\ln 95 - \ln\ 5 + (0.05)(1)}{0.05} = 60\ \text{m}^2$$

Therefore, if the frequency of a randomly distributed species is 5% by measurement with a small plot, then measurement with a plot 60 times the size of a smaller plot gives a frequency of 95% for the same species. Hyder *et al.* (1965) computed the theoretical lower limits in the size of complementary large plots for measurement of frequency of a species greater than 85% when its frequency is 5% and 10% in a small plot.

Hyder *et al.* (1965) did trial-and-error sampling with plot sizes 1652, 523, and 26 cm^2 to find one that sampled blue grama (*Bouteloua gracilis*) in a frequency range of 63–86%. The plot size of 26 cm^2 gave an adequate estimate of frequency for blue grama and only four other species. So, to obtain an adequate estimate of frequency for other species, the need for a complementary plot existed. We have already seen that to measure frequency in the range between 5 and 95%, the larger plot should be 60 times the size of the smaller plot. The plot size closest to 60 times 26 cm^2 was 1652 cm^2.

In contrast to finding plot sizes to be used in the measurement of frequency, Helm (2010) also used the Poisson distribution to convert the frequency per plot size sampled to an expected frequency for any other plot size. Hyder *et al.* (1965) estimated the second plot size from the one sampled, while the object in Helm's study was to allow a comparison to be made of frequency results from different plot sizes. Helm used an example of frequency data from a sagebrush-bunchgrass vegetation type in Oregon, USA and derived a single equation to estimate expected frequencies as they vary by plot sizes. This equation is

$$F_e = 1 - (1 - F_s)^r \text{ and applies when } 0 < F_s < 1 \quad (5.10)$$

where

F_e is the estimated frequency
F_s = measured frequency from a sampled plot size
r = A_e/A_s
A_e = plot size of interest
A_s = the plot size sampled

The assumption is made that plants follow an approximately random distribution that is assumed to be described by the Poisson distribution. In most cases the estimated frequency F_e should be sufficient for practical applications. It is worthwhile to note that a problem could result in the estimated frequency value of a species if the density of that species is different in the plant communities being compared.

Sampling designs for field measures of frequency can be accomplished either by random or non-aligned systematic location of plots in the area to be sampled. It is important to sample the entire area covered by the plant community to mitigate patterns in species aggregations that are present. Furthermore, one can select random or non-aligned systematic points for placement along randomly or non-aligned systematically located line transects (Figure 4.2).

It has been suggested that frequency is not an absolute measure because it is a function of size and shape of the sample unit (Greig-Smith 1983). There were attempts made to obtain an absolute measure by eliminating the effect of the size of the sample unit. This could be accomplished by reducing the size of the plot area until only a point exists, then using point methods to estimate frequency. In such case, the measure of frequency no longer exists because the frequency measure is defined according to the plot area in which the plant occurred.

There are two commonly used arrangements of nested plots. One is based on an arrangement of sub-plots into a given corner of successively large plots (primary plot); i.e., square or rectangular plots (Figure 2.9). The other arrangement consists of either concentric square, rectangular, or circular plots that are nested on a geometrical basis; i.e., all successive plots have the same identical center location within the primary plot (geometrical basis) (Figure 5.1b).

Outhred (1984) had introduced two alternative procedures for estimating root frequency and named the resulting estimates, frequency scores and importance scores. The intent of sampling was to eliminate the practical limitations of the standard frequency estimates; i.e., dependency on size of sub-quadrats used and on the spatial patterns of plants within the area sampled. Later Morrison *et al.* (1995) recounted the development and use of square, concentric plots with a common center of the primary plot, to obtain frequency estimates as reported by Outhred (1984). Morrison and his colleagues conducted an assessment of these techniques and found that the methods used to estimate frequency, as proposed by Outhred were less affected by the choice of sub-quadrat size and the spatial patterns of plant species. These methods are easily employed in the field (Figure 5.1a, b).

Using Figure 5.1a, assume 25 points are located to represent x, y coordinates of individual plants. Let us assume that each point represents an individual plant of species Y. Furthermore, assume that the large plot is a 10×10 m plot with a nested sub-plot of 4×4 m plot and within this latter plot, six of the 25 individual plants occurred. Finally, nine contiguous 1×1 m, (1 m^2) sub-plots are placed within the 16 m^2. These sub-plots are used to estimate frequency at a scale of 1 m^2 area; i.e., each 1 m^2 plot is examined for the presence of species Y. A reading of these plots show that four contained one or more individual plants. Then from Equation 5.1, the frequency count of the species was $F_c = 4$ (number of plots that contained at least one individual of the species). Furthermore, the nearest area scale of the frequency proportion was 16 m^2 with $F_p = 4/9 = 0.44$, from Equation 5.3.

Figure 5.1b illustrates how density, dispersion and quadrat size affects frequency estimates. Consider that estimates for the frequency score and the importance score are obtained from a compounded quadrat of seven nested sub-quadrats. In this case, the frequency score is the total number of nested sub-quadrats in which at least one individual plant occurs; i.e., Species Y occurred in four sub-quadrats and the proportion is $4/5 = 0.80$. Because the sub-quadrats have areas that increase geometrically, frequency scores have a logarithmic relationship to density, as does the standard frequency measure. Additionally, areas of the sub-quadrats are unequal and frequency scores include patterns that occur over the concentric quadrat sizes compared to standard frequency values that are based on equal-area sub-quadrats.

In the example, beginning with the smallest plot, each sub-quadrat is searched for the occurrence of the species up to largest plot. If the species is found in the smallest sub-quadrat, the species is assigned the maximum frequency score of 7. This is because the species automatically occurs in all seven sub-quadrats. As the species is encountered in a successively larger quadrat, the importance score i becomes smaller, $i = 1$ for the largest quadrat to 7 for the smallest quadrat in the example. Note that the concentrically placed sub-quadrat areas do not include the area of the smaller sub-quadrats, whereas the "nested plots" described by Hyder *et al.* (1965) have accumulative areas made up of successively smaller plots.

In practice, a species list is made for each sub-quadrat and the frequency score is given by counting the number of sub-quadrats containing the species. Only newly occurring species in the next sized quadrat are listed, progressing from the smallest to the largest sub-quadrat and the importance score for the species is given by the quadrat in which it is listed. Furthermore, the importance score of a species is an indication of the size of quadrat needed to measure the frequency of that species. Morrison *et al.* (1995) point out that the score given the individual species is an inverse estimate of the area occupied by the plant (MA), i.e., a density estimate $= \frac{1}{MA}$.

Table 5.1 Commonly used quadrat sizes (cm) for recording nested rooted frequency in plant communities by the United States Forest Service (metric dimensions). (Reproduced from Sampling Method. Caratti, J. 2006. The United States Forest Service. *Gen. Tech. Rep.* RMRS-GTR-164-CD. 15 pp.)

NRF numbers	Standard	Grassland communities	Sagebrush communities	Pinyon-juniper
Sub-plot 1	5 × 5	5 × 5		
Sub-plot 2	25 × 25	5 × 5	10 × 10	20 × 20
Sub-plot 3	25 × 50	10 × 10	20 × 20	50 × 50
Sub-plot 4	50 × 50	20 × 20	50 × 50	100 × 100

Percentage area of a 50 × 50 cm plot	
5 × 5 cm	1%
25 × 25 cm	25%
25 × 50 cm	50%
50 × 50 cm	100%

Caratti (2006) has recommended a number of quadrat sizes for sampling of species frequency (Table 5.1). These plot sizes were selected after sampling various vegetation types. A combination of these plots can be arranged into nested plots (Figure 5.1) or in concentric squares.

Sampling designs for field measures of species frequency can be accomplished either by random or non-aligned systematic location of plots in the area to be sampled. It is important to sample the entire area covered by the plant community to avoid patterns in individual species aggregations that are present. Furthermore, one can select random or non-aligned systematic points for plot placement along randomly or non-aligned systematically located line transects.

Frequency data converted to proportions often follow the Poisson or point binomial distribution. Therefore, mean and variances are estimated differently from those of normally distributed populations. Species frequency, as a proportion, is estimated from a random sample of plots as Equation (5.3) shows (repeated here for convenience)

$$p = \frac{\text{number of quadrats occupied}}{\text{total number of quadrats examined}} \tag{5.11}$$

where p is the estimated frequency proportion (F_p) as given in Equation (5.3). The mean of a frequency estimate as a proportion (p) from randomly located sampling units is estimated from Equation (5.11) while the variance is estimated as

$$\text{variance} = pq \tag{5.12}$$

where p is the estimated mean frequency proportion of a species and q is the complement $(1-p)$ (Cochran 1977).

A combination of plots and transects is used to obtain an estimate of the variance components of plots s_{qd}^2 and transects, s_t^2 estimated from

$$s_{qd}^2 = \frac{k\left(\sum_{i=1}^{n} p_i q_i\right)}{n(k-1)} \tag{5.13}$$

$$s_t^2 = \frac{\sum_{i=1}^{n} (p_i - \bar{p})^2}{n-1} - \frac{s_{qd}^2}{k} \tag{5.14}$$

where k is the number of plots per transect (constant from transect to transect), n is the number of transects, and p and q are the same as before (Hyder *et al.* 1965).

Optimum sampling depends on both the cost and the variance of the frequency data, and there are two components of cost: (1) time involved in locating the observations (C_t) and (2) recording data at each measurement location (C_{qd}). The total time required to record species presence in a plot is negligible compared to the total time required to locate observations. Consequently, to economize the cost of frequency measurements, many ecologists prefer to use a sample design in which plots are located along line transects. In which case, one should use non-aligned systematically located transects. The process is a two-stage sampling design. Generally, the number of plots is constant from transect to transect.

The optimum number of plots k per transect is given by the relationship

$$k = \sqrt{\frac{s_{qd}^2 C_t}{s_t^2 C_{qd}}} \tag{5.15}$$

The optimum number of transects n is estimated as

$$n = \frac{4nV_p}{\left(2\sqrt{\frac{pq}{k-1}}\right)} \tag{5.16}$$

where n is the number of transects used to obtain a preliminary sample and k is the number of plots per transect, and $4 = t^2$ where $t \approx 2.0$ from the t-table for a 0.95 confidence limit. An efficient sample plan requires computation of k and n for each species. One can then use either the maximum value of k and n or select k and n such that a maximum number of the most important species are adequately measured. The computation for optimum sampling is a lengthy and tedious process. Therefore, for all practical purposes, 4 plots randomly located on 25 randomly placed transects on a systematic grid should give satisfactory results within a homogeneous plant community.

5.5 Applications for frequency data

Applications for the frequency measure include the inventory of plant communities to determine species composition, prediction of biomass by species, and monitoring frequency changes in plant species over time. Changes in species frequency are determined for effects of livestock or big game grazing, detection of changes in a threatened–endangered plant species, or as a measure of successful restoration of a disturbed natural area.

5.5.1 Prediction and monitoring

Frequency data can be combined with cover data to study and compare species changes in plant communities. An in-depth interpretation of results was attained from such a study in Sweden (Bergfur *et al.* 2004). These researchers used measures of both plant species frequency and cover to describe and interpret species seasonal phenological changes over a growth season in two semi-natural grasslands grazed by cattle. Data from the two sites were analyzed separately because of site differences. The xeric grassland had been grazed for several hundreds of years, while the other grassland was mesic and grazed more often in recent times. Species frequency and cover were measured at four times during the growing season for the grasslands, using eight 0.5 m^2 plots that were divided into 25 squares each. Presence/absence data for species were recorded in each sub-plot, while cover was estimated over the whole plot. Temporal patterns of species phenological changes, over the growth season, were interpreted to illustrate the importance of timing field sampling to make valid comparisons of plant communities. Variations in phenological changes were accounted for by changes in abundances of species over time, not by appearances/disappearances of species.

Römermann *et al.* (2007) studied species frequencies to predict species habitat occurrence on a grid-cell basis. They noted that a species with a low frequency value is not necessarily rare if it occupies all suitable habitats. This would be born out by these species being important predictors of their own limited habitats. Prediction of habitat frequency distribution for the study species depends upon plant species frequency data availability on a large-scale basis. The study was carried out over Germany and Bavaria. Based on readily available data banks for these countries, it was possible to predict habitat distribution and frequency from the co-occurrence of habitat-specific species per grid cell. Habitat distribution and frequencies were retrieved at the national scale on the basis of habitat-specific species co-occurrences per grid cell. Species distribution data only was used successfully in formulation of models that predicted frequencies of 24 habitats based on the co-occurrence of 24% of formation-specific species per grid cell. This study showed that species habitat distributions and their frequencies can be deduced from the co-occurrence of habitat-specific species.

A reason for collecting frequency data could be to combine it with one or more of the other measures to study species composition changes over time. This measure is efficient in detecting changes quickly and with the other measure(s) enable an interpretation to be made of the change within the community; for instance, the effects of grazing by livestock on a plant community of interest may appear first in the frequency of a species and indicates that distribution of the species is changing in its distribution over the community. Hironaka (1985) found that frequency of species does indicate that a change in vegetation has occurred. He then demonstrated that on most study sites frequency data are capable of detecting such a change more rapidly and cost less than the other measures (cover, density, and biomass).

West (1985) concurred that frequency does provide the most easily obtained early warning of changes in species; this is especially true for key or indicator species, as well as for invasive species. West (1985) issued a warning that a greater risk of error using frequency data is made by concluding that a downward trend in species is occurring in the plant community when in fact it is not doing so. He attributed such an error in change of frequency trend to be either upward or downward because the sample size was inadequate. That is, the area of the plant community sampled has to include the full extent of the presence of the species within the community. West (1985), like other plant ecologists, also expressed concern about frequency data values being dependent on plot size as well as on the spatial distribution of plants within a plant community. However, both of these effects of frequency data can be mitigated by a sampling design for quadrats, as given in Figure 5.1 above.

Mosley *et al.* (1987) evaluated frequency–biomass relationships to determine the feasibility of predicting individual and total biomass of plant species from their frequency values. Plant species herbaceous yield and frequency data were jointly obtained from dry mountain meadows sites in central Idaho, USA. The frequency data were used to predict species biomass. These data were obtained from a macro-plot, 100 × 100 ft (30.5 × 30.5 m), sampled by five parallel 100 ft (30.5 m) transects placed 25 ft (7.6 m) apart. Rooted presence of each species was recorded in quadrats, 10 × 10 cm, placed at 5 ft (1.5 m) intervals along the five transects; therefore 100 frequency quadrats were measured per site.

Biomass estimates were obtained from the same macro-plot above by use of 10 circular plots, 9.6 ft^2 (0.29 m$^{2)}$ placed at 3.05 m intervals along each of three of the five 100 ft (30.5 m) transect lines described above. Each of the 30 plots was clipped and biomass determined according to species. Herbage biomass data was used to calculate and monitor carrying capacity for livestock grazing of the area.

The relationship between estimated and actual total site yield was found to be linear when frequency by species was summed and used to predict total biomass by species. Prediction models overestimated yield at the lower range of values for biomass (<800 kg/ha) and underestimated actual yield at the higher biomass levels.

5.6 Considerations

Frequency is defined as the occurrence of a plant within a sampling unit area such as a 0.1 m^2. If sampling is conducted by walking through a plant community and plant species are observed, presence of a species can be recorded rather than a frequency value from quadrat sampling. This practice has been found to support the adage that vegetation is an expression of its environment. This being the case, the presence of plant species is important for the prediction of habitat or the extent of the species distribution throughout a country or the world. Elith *et al.* (2006) conducted an intensive study of the use of plant species presence data to predict species distributions. They compared 16 modeling methods that have been used over time and acquired presence data for 226 species for six regions of the world to test models. Their study was indeed the most comprehensive set of model comparisons to date (as of study year 2006) for evaluation of prediction of species distribution models. Although presence-only data were used in these models, these predictions were evaluated with independent presence–absence data for species. This study showed that presence-only data were found to be effective for modeling species distributions over a wide range of locations around the world. Furthermore, their "novel" methods, as they referred to them, consistently outperformed more established techniques. The importance of such a study enables the use of plant species records that reside in museums and herbaria around the world.

Plot size and shape are often thought by some vegetation ecologists to make the frequency measure difficult to interpret. To eliminate this problem in the past, the area of a plot in mathematical theory was allowed to become smaller and approach zero (0). Then the plot resulted in an infinitesimally small area shaped as a square, rectangle or circle depending on the shape of an assumed original plot. This approach to deal with the effect of plot size on frequency results in a contradiction. By the strict definition of frequency, a contact on any part of a plant has to be associated with a defined area (plot) of ground beneath that contact. This area has to be finite, which violates the reason for obtaining such a small area in the first place, i.e., to obtain a measure of frequency independent of the plot area. Therefore, a contact by point without an area being associated with it cannot be a frequency value (0, 1) that is measured. Furthermore, the plant contacted has to be "rooted" within the plot because frequency is not independent of the density measure. Contact by a point on an aerial part of a plant does not require this to happen.

The use of point lines or frames located within a plot to obtain contacts on plants does not qualify as frequency data because each contact is not independently defined either by volume or area of ground. Otherwise, the effects of plot size, plant size, and individual plant species distributions all affect the measure of frequency (Archibald 1952, Aberdeen 1958, Greig-Smith 1983). Intensive plot or quadrat sampling with a size that suits the objectives of the field ecologist will yield unbiased estimates of frequency. The requirement is that the area covered by the population has to be sampled regardless of sample size adequacy. One might want to consider the use of a non-aligned systematic sample design.

5.7 Bibliography

Aberdeen J. E. C. 1958. The effect of quadrat size, plant size, and plant distribution on frequency estimates in plant ecology. *Aus. J. Bot.* **6**: 47–58.

Archibald E. E. A. 1952. A possible method for estimating the area covered by the basal parts of plants. *S. Afr. J. Sci.* **48**: 286–292.

Bergfur J., Carlsson A., and Milberg P. 2004. Phenological changes within a growth season in two semi-natural pastures in southern Sweden. *Ann. Bot. Fennici* **41**: 15–25.

Bonham C. and Clark D. 2005. Quantification of plant cover estimates. *Grassl. Sci.* **51**: 129–137.

Brown D. 1954. *Methods of Surveying and Measuring Vegetation.* Bulletin No. 42, Commonwealth Bureau of Pastures and Field Crops: Hurley, Berkshire.

Cain S. A. and Castro G. M. 1957. *Manual of Vegetation Analysis.* Harper: New York.

Caratti J. 2006. *Cover/Frequency (CF) Sampling Method.* United States Forest Service General Technical Report RMRS-GTR-164-CD.

Chen J., Shiyomi M., Bonham C., Yasuda T., Hori Y., Yamamura Y. 2007. *Cover Estimation Based on Beta Distribution.* Ecological Research, Japanese Grassland Society.

Cochran W. G. 1977. *Sampling Techniques.* John Wiley & Sons, Inc.: New York.

Elith J., Graham C. H., Anderson R. P., Dudik M., Ferrier S., Guisan A., Hijamans R. J., Huettmann F., Leathwick J. R., Lehmann A., Li J., Lohmann L. G., Loiselle B. A., Mannion G., Moritz C., Nakamura M., Nakazawa Y., Overton J. McC., Peterson A. T., Phillips S. J., Williams S., Wisz M. S., and Zimmermann N. E. 2006. Novel methods improve prediction of species' distributions from occurrence data. *Ecography* **29**: 129–151.

Greig-Smith P. 1983. *Quantitative Plant Ecology*, 3rd edn. University of California Press: Berkley, CA.

Helm S. 2010. A calculation of expected plant frequency. *Castanea* **75**: 226–231.

Hironaka M. 1985. Frequency approaches to monitor rangeland vegetation. *Proceedings of the 38th Annual Meeting of the Society for Range Management, February 1985, Salt Lake City.* Society for Range Management, 87–90.

Hyder D., Bement R., Remmenga E., and Terwilliger C., Jr. 1965. Frequency sampling of blue grama range. *J. Range Manage.* **18**: 90–94.

Jongman R., ter Braak C., and van Togeren O. (eds) 1995. *Data Analysis in Community and Landscape Ecology*. Cambridge University Press: New York.

Mayland H. F. 1972. Correlation of the exposure and potential solar radiation to plant frequency of *Agropyron desertorum. Ecology* **53**: 1204–1206.

Morrison D. A., Le Brocque A. F., and Clark P. J. 1995. An assessment of some improved techniques for estimating the abundance (frequency) of sedentary organisms. *Vegetatio*, **120**: 131–145.

Mosley J. C., Wester D. B., and Bunting S.C.. 1987. Estimating herbaceous yield from species frequency. *Northwest Sci.* **61**: 55–59.

Outhred R. K. 1984. Semi-quantitative sampling in vegetation survey. In: Myers K., Margures C. R., and Musto I. (eds), *Survey Methods for Nature Conservation.* CSIRO: Canberra, 87–100.

Raunkiaer C. 1909–10. Formationsundersogelse og formationsstatistik. *Botanisk Tidsskrift* **30**: 20–132 (English translation Raunkiaer 1934 below).

Raunkiaer C. 1934. The life forms of plants and statistical plant geography (English translation of collected papers, see pages 201–282). Clarendon Press: Oxford, UK.

Reich R., Aguire-Bravo E., and Briseno M. 2008. An innovative approach to inventory and monitoring of natural resources in the Mexican State of Jalisco. *Environ. Monitor. Assess.* **146**: 383–396.

Romell L. G. 1930. Comments on Raunkiaer's and similar methods of vegetation analysis and the "law of frequency." *Ecology* **11**: 589–596.

Römermann C., Tackenberg O., Scheuerer M., May R., and Poschlod P. 2007. Predicting habitat distribution and frequency from plant species co-occurrence data. *J. Biogeography* **34**: 1041–1052.

Shiyomi M., Gaborcik N., Koizumi H., Javorkova A., Uhliarova E., and Jezikova O. 2004. Spatial patterns and species diversity of plant communities in sown, oversown, and semi-natural grasslands in Banska Bystrica, Slovakia. *Grassl. Sci.* **50**: 1–8.

USDA 1999. *Sampling Vegetation Attributes*. Interagency Technical Reference. US Department of Agriculture, 1734-4. Natural Business Center: Denver CO.

West N. 1985. Shortcomings of plant frequency-based methods for range condition and trend. Proceedomgs of the 38th Annual Meeting of the Society for Range Management, Febraury 1985, Salt Lake City. Society for Range Management, 87–90.

White G. C. and Bennetts R. E. 1996. Analysis of frequency count data using the negative binomial distribution. *Ecology* **77**: 2549–2557.

6

Cover

Cover of vegetation is the percentage of ground surface covered by vegetation material. Other definitions exist, but it may be difficult to obtain a measure with respect to the definition. For example, Daubenmire (1959) suggested that cover is an approximation of the area over which a plant exerts its influence on other parts of the ecosystem and is not an estimate of the shaded area on the ground. This definition led to the approximating polygons of plant canopies that occurred within his plot rather than on larger areas used by European ecologists. Daubenmire's definition of cover never became widely accepted by vegetation ecologists, rather they have used various plot sizes and shapes to estimate ground area covered by plants. The plot used by Daubenmire, however, is probably the most used and best known among plant ecologists. It is noted that all historical cover measurements, for example the methods of Raunkiaer (1934), Braun-Blanquet (1965), Tansley (1947), and Daubenmire (1959, 1968), were originally obtained to describe plant communities, not to make statistical comparisons among these communities, because these cover measurements were primarily based on a qualitative approach using scales, not absolute values of cover. In time, all scales were assigned numerical values to conduct statistical analyses.

Some definitions of cover are:

- Vegetation cover: total cover of vegetation on an area of ground.
- Crown cover: canopy of trees and shrubs projected onto a ground area.
- Ground cover: cover by plants, litter, and rocks in a vegetation type.
- Range plant cover: cover of all plants available to livestock and other herbivores.
- Habitat cover: cover of vegetation as viewed horizontally for wildlife protection.

A characteristic that is common in all the above, and other definitions, is that cover is always the vertical or horizontal projection of vegetation parts, although individual leaves occur at all angles. Cover is expressed as a proportion, percentage, or amount of cover from a scale basis (e.g., ranging from 1 to 5).

Measurements for Terrestrial Vegetation, Second Edition. Charles D. Bonham.

When cover of a species or plant life-form is expressed as a percentage of total vegetation, it is referred to as "relative" cover of the species and when summed over species, equals 100%. Cover is one of the more commonly measured quantities in vegetation sampling. One major advantage of cover as a quantitative measure is that different plant life forms (e.g., mosses, forbs, grasses, shrubs, trees) can all be evaluated in comparable terms. If the vegetation has a distinct layered structure (e.g., trees, shrubs and undergrowth), then the cover of species in each layer is measured separately.

Cover is usually less than total leaf area because many leaves overlap each other. Often there are open spaces in foliage, and in that case, cover will actually be less than total crown or shoot area. Many plants have dissected clumps, and there are gaps in plant canopies. Dissected clumps should be measured as individual clumps and the area of dead plant center should be subtracted from the area of the entire clump.

6.1 Basal area techniques

The area outline of a plant near the ground surface is referred to as "basal area." This term is widely used in silviculture and rangeland management and refers to the total stump surface area of trees at breast height (1.4 m) or the area of ground covered by basal parts of grasses. Basal area measurements are widely used on bunchgrass or tussock vegetation, and measurements are recorded at a height of approximately 2 cm. Then foliage or basal area cover of species falling within a plot is measured, or visually estimated, and recorded. This method is especially suitable for vegetation types composed of species that have well-defined clumps, bunches, tufts, or tussocks. Usually, the choice of size and shape of the quadrat for these measurements depends on the type of vegetation and objectives for measurements. However, quadrats of areas 1, 4, and 10 m^2 are commonly used, respectively, for grassland, shrub, and forests, either alone or in nested form for the specific vegetation type. A nested design is used when plant species from these general types occur together in combinations within a plant community (Figure 2.9). However, the nested plot design should be updated to that shown in Figure 5.1 so that each plot size has a common center for each larger plot.

Measurement of basal area may be more reliable for estimation of cover changes for some species than foliage cover because basal cover fluctuates less during seasonal periods of precipitation or other perturbations. However, in shrubs, forbs, and single-stemmed grasses, basal area may not the best measure because of the number of small stems to be measured in a sample unit. Basal area measurements of perennial plants have practical application on permanent plots where vegetation changes are to be monitored over time.

6.1.1 Diameter rule

Pearse (1935) used the diameter rule method (Figure 6.1) to measure the cover of forbs and grasses. This method is simple to construct and is easily used in the field.

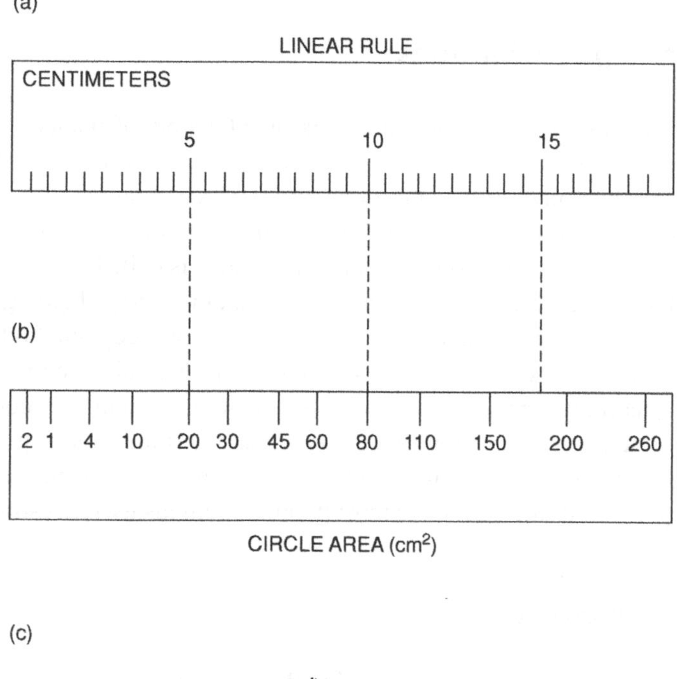

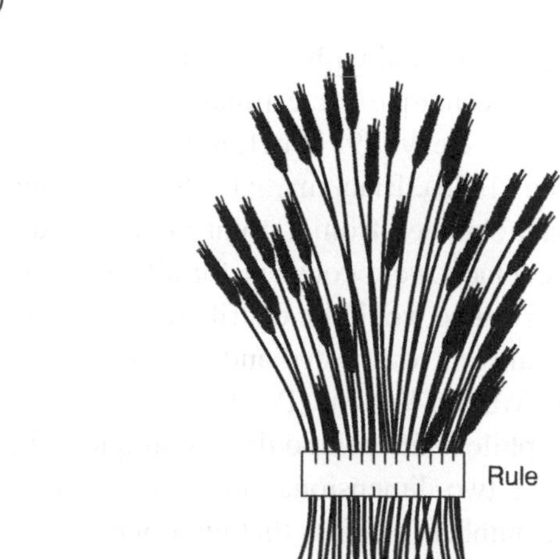

Figure 6.1 Linear-diameter measure for estimation of basal area of plant (Republished with permission of Ecological Society of America, from An area-list method of measuring range plant populations. Pearse, K. 1935. *Ecology* **16**: 573–579; permission conveyed through Copyright Clearance Center, Inc.)

Pearse used a ruler marked with a special scale from which the area of a circle could be determined by measuring its diameter. Each plant measured is compressed with the hand until ground surface is not visible through the foliage, and the area of the clump is determined with the area ruler. Plots can be used as a way to measure individual plants.

6.2 Intercept techniques

Intercept techniques use linear measurements of intercepts of plants by lines and points. These intercepts can be made by a length along a horizontal line, the number of contacts made by points with plants, or counting of ocular sightings of plant parts with a system of cross-hairs, grid points, or dot matrices on a transparent surface. The use of a line to estimate cover by intercept distance is called the "line intercept transect method" (Figure 2.3). In comparison, a system of cross hairs, grid points, or dot matrices are grouped under the heading of "point intercept methods." In fact, both lines and points are plots. The line point is the logical outcome of a two-dimensional quadrat when its size is reduced in one direction until it becomes a line. Similarly, a point quadrat is the logical outcome of a two-dimensional quadrat when its size is reduced in both directions until it is reduced to a point. Line intercept and point intercept methods are two of the most popular methods used to estimate cover.

6.2.1 Point intercept

Technically, a point has an area associated with it, although the area is very small, and therefore is a quadrat. The technique for estimates of cover by points in a frame was used in New Zealand by Levy (1927) before the method was more fully described by Levy and Madden (1933). Blackman (1935) contributed much to the statistical characteristics of data obtained from a point method and suggested that if the expected proportion of contacts were constant for all locations of the frame, then the number of contacts per frame would be distributed as a positive binomial. Subsequently, Blackman's initial work was tested and improved by Goodall (1952, 1953), Winkworth (1955), and Warren-Wilson (1959a,b, 1960, 1963a,b) for use in different vegetation types. The philosophy behind this technique is that if an infinite number of points are placed in a two-dimensional area, exact cover of a plant can be determined by counting the number of points that hit a plant.

The technique, in its usual form, is more suited to herbaceous vegetation such as prairie and grazed pastures. The point method provides estimates of: (1) percentage ground cover by species, (2) percentage cover each species contributes to the total area, and (3) percentage that each species contributes to a vegetation type (number of hits per species per number of vegetation hits).

6.2.2 Grid-quadrat frame

A quadrat gridded by cross-line intersections that are treated as intercepts by individual points has been used for several decades (Cook and Bonham 1977). More recently, a grid-quadrat frame described by Shyomi and Yoshmura (2000) results from the division of a quadrat into n smaller quadrats of equal size. Such a grid-quadrat frame can be of any shape and size. Cross-points of a grid quadrat are, in fact, points. Any vertical interceptions of cross-points of a grid with plant parts are considered hits. Plant ecologists have reported that the angle of viewing cross-points introduces error, and two persons viewing at different angles obtain different cover values. This problem can, however, be overcome by having a double grid of cross-points, which greatly reduces the possibility of different observers aiming at different points.

Chen *et al.* (2008) used a transparent grid plate that had 100 points arranged in a lattice with 1 cm grid spacing (10 × 10 cm); one grid point count represents 1 cm^2 of cover. Several previous studies had shown that data from single points, point-frame, and the point-grid cover methods can be analyzed by the beta distribution to estimate cover (Hughes and Madden 1993; Shyomi and Yoshmura 2000; Shyomi *et al.* 2004; Chen *et al.* 2006; Huang *et al.* 2007). Details are provided by Shyomi and Yoshmura (2000) and Shyomi *et al.* (2004) for derivation of the beta distribution from the beta-binomial distribution. The probability density function of the beta distribution with parameters ν and ω, is expressed as

$$f(\nu, \omega) = \frac{x^{\nu-1}(1-x)^{\omega-1}}{B(\nu, \omega)} \tag{6.1}$$

where $B(\nu,\omega)$ is the beta function given by

$$B(\nu, \omega) = \int_0^1 u^{\nu-1}(1-u)^{\omega-1} \mathrm{d}u \tag{6.2}$$

Kemp and Kemp (1956), Goel and Strebel (1984), and Chen *et al.* (2008), among others, discussed the analysis of point data by use of the beta distribution. Goel and Strebel (1984), in particular, used the beta distribution to describe the leaf orientation in plant canopies, while Chen *et al.* (2008) showed the beta distribution to have wide application in sampling designs and sample size, as well as statistical comparisons for estimates of cover from point data. The moment and maximum likelihood equations for parameters of the beta distribution are

$$\text{Mean}(\mu) = \frac{\nu}{\nu + \omega} \tag{6.3}$$

An estimate of ν is: $\bar{x}([\bar{x}\,(1 - \bar{x}\,)/s^2] - 1)$.

$$\text{Variance } (\sigma^2) = \frac{\nu\omega}{(\nu + \omega)^2\,(\nu + \omega + 1)} \tag{6.4}$$

An estimate of ω is: $(1 - \bar{x})\{[\bar{x}(1 - \bar{x})/s^2] - 1\}$, where $\bar{x}$ = sample mean and s^2 = sample variance (unadjusted) (i.e., $\frac{\sum x^2}{n}$, not $n - 1$ when $x = x - \bar{x}$ (Hastings and Peacock 1975). Use these estimates to find the mean (μ) (Equation 6.3) and variance (σ^2) (Equation 6.4) for the beta distribution where μ and σ^2 are the mean and variance for the beta distribution and ω is the number of hits or contacts by pins on plant parts in a given sampling unit. The mean number of contacts for kth frame is n_k, where n is the number of frame placements. The mean is a measure of the overall expected proportion, and the variance of the beta distribution is the amount of variation about this overall proportion from location to location (Kemp and Kemp 1956). The variance (v) of number of contacts per frame is

$$v = n_k(1 - \mu) + n(n - 1)\sigma^2 \tag{6.5}$$

if the expected proportion of contacts had this variability from frame to frame. Then, $n(n - 1)\,\sigma^2$ measures the variance in excess of that expected on a positive binomial assumption where a constant proportion is expected.

The total number of pin observations needed, then, is reduced if the number of pins per frame is decreased. However, the number of locations would probably be increased. The variance of percentage cover value estimated from N_n frames of n pins is

$$V = \frac{\mu\,(1 - \mu)}{N_n n}\left(\frac{\nu + m + n}{\nu + m + 1}\right) \times 10^4 \tag{6.6}$$

Substitute frame by grid-quadrats when the latter are used to obtain hits. Patchiness in species cover can be studied if there are at least two pins used per frame (or grid-quadrats) so that intraframe variance can be compared with interframe variance. That is, results from Equation (6.6) compared to those from Equation (6.5).

It is noteworthy that Chen *et al.* (2008), in particular, showed that as $n \to \infty$ and $\mu \to \infty$ under the condition of μ/n = constant, the beta-binomial distribution asymptotically approaches the beta distribution (Equation 6.1), and $x = n_i/n$ asymptotically approaches cover ($0 \leq x \leq 1$). From these conditions we can assume that cover is a continuous variable that follows the beta distribution (Chen *et al.* 2007). This density distribution is used to estimate the probability (P_i) for number of contacts ($i = 0, 1, \ldots, n$) made by a pin or appears at n intersections on a grid plate.

6.2.3 Vertical point frame

A vertical point frame (Figure 2.1) is a much-used, practical tool to obtain estimates of cover. It consists of a metallic frame with two legs and two cross-arms (Levy and Madden 1933). The two cross-arms have 10 or more perpendicular equidistant holes through which steel rods or wire pins are slid and hits are recorded by species. One can either use the same pin for 10 observations or use a separate pin for each hole. However, experience has shown that working with a single pin is slower than having 10 pins. In effect, the point frame is the same as a line-point transect, which has greater length and more points. The size of the frame depends on vegetation height and patterns that, in turn, affect the spacing of pins and height of the frame.

6.2.4 Horizontal point frame

Flippin-Dudley (1996) developed a horizontal point frame (HPF) (Figure 6.2) to obtain estimates of stream flow resistance caused by vegetative drag. Dudley *et al.* (1998b) then reported on a specific study of the application of the frame and showed that an estimate of vegetation density ($\mathrm{Veg_d}$) was obtained from the number of horizontal hits on vegetation. In particular,

$$\mathrm{Veg_d} = \frac{\text{Number of hits}}{\text{Number of points}} \times \frac{1}{D} \tag{6.7}$$

where D is the length of the pin used (0.3 m). Drag depends on the density of the vegetation, $\mathrm{Veg_d}$, which is interpreted as the frontal area of submerged vegetation projected onto a plane that is perpendicular to the direction of flow per unit volume of flow. Equation (6.8) provides an estimate of this frontal area in a similar way that vertically placed pins in a frame provides an estimate of proportion of ground covered by vegetation (Flippin-Dudley 1996). Dudley *et al.* 1998a) used pins to obtain the number of contacts on vegetation along a horizontal line-of-sight. However, contacts on all vegetation along the route of these pins are recorded because each vegetation contact contributes to resistance of water flow in the stream. Data would resemble recording all contacts along a vertical route to ground level for the purpose of estimating species composition for a plant community. Note that contacts along the routes of pins create a three-dimensional space (volume) by using rows of pins that are spaced at selected vertical intervals of height above the stream bed (Figure 6.2). The result of Equation (6.7), $\mathrm{Veg_d}$, is used to estimate vegetative drag from

$$\text{Vegetation drag} = \mathrm{Veg_d}\frac{\sum A_i}{AL} \tag{6.8}$$

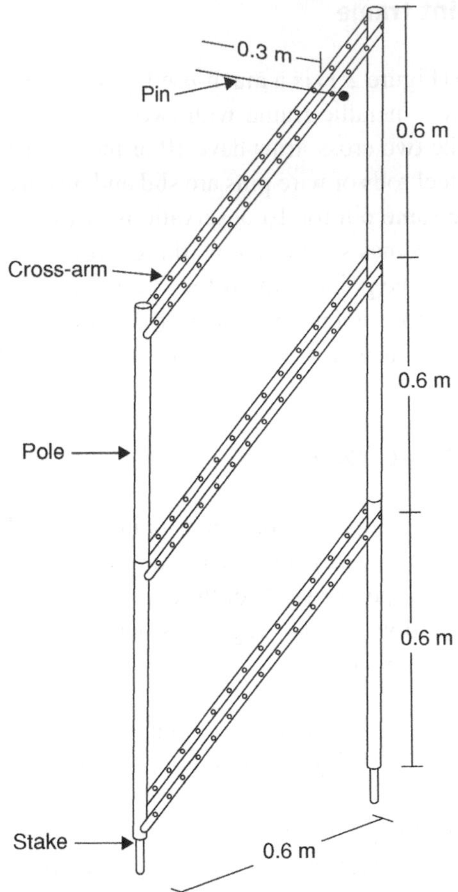

Figure 6.2 Schematic drawing of HPF with three sets of cross-arms, each fabricated with 20 holes used for guiding pins horizontally through vegetation. (Reproduced with permission from Comparison of methods for measuring woody riparian vegetation density. Dudley *et al.* 1998. *Journal of Arid Environments* **38**: 77–86.)

where ΣA_i is the sum of the frontal areas of the *i*th vegetal elements projected onto a plane perpendicular to the direction of flow, A is the cross-sectional area of flow, and L is the length of the channel investigated. Other methods to estimate vegetation density include the board method (MacArthur and MacArthur (1961) and a camera method (MacArthur and Horn 1969). The three methods were compared by the Multi-Response Permutation Procedure (MRPP) developed by Miekle (1986). Dudley *et al.* (1998a) used the procedure to test the relationship among the Veg_d data from the three methods of estimation. All two-way comparisons of the three methods differed significantly from one another for the two stream locations (Tables 6.1 and 6.2). Table 6.3 shows that sample adequacy size was considerably

Table 6.1 Mean Veg_d values associated with the three methods and percentage difference from HPF values. (Reproduced with permission from Dudley, S., Bonham, C., Abt, S., and Fischenich, J. 1998. Comparison of methods for measuring woody riparian vegetation density. *J. Arid Environ.* **38**: 77–86. © Academic Press.)

Site	Method	Total no. measurements	Mean Veg_d (m^{-1})	Difference from HPF (%)
Spring Creek	HPF	76	0.279	
	Board	50	0.259	7
	Camera	34	0.177	36
Poudre River	HPF	50	0.353	
	Board	50	0.253	28
	Camera	50	0.253	28

greater for HPF compared to the other two methods and, thus, costs were greater for this method.

6.2.5 Single points

A given number of individual points, randomly distributed, can give a more precise estimate of cover (Table 6.4) than if the same number of points is grouped into frames (Blackman 1935, Goodall 1952, Greig-Smith 1983). Single-pin measurements require one-third as many points as required when groups of pins are used. Comparable accuracy (Goodall 1952) and time required is reduced from one-sixth to one-eighth of that required for the point-frame method (Evans and Love 1957). This reduction occurs because the grouping of pins results in overestimation of cover owing to spatial auto-correlation among the hits on the same species. A

Table 6.2 Relationship among data for three methods using MRPP. (Reproduced with permission from Dudley, S., Bonham, C., Abt, S., and Fischenich, J. 1998. Comparison of methods for measuring woody riparian vegetation density. *J. Arid Environ.* **38**: 77–86. © Academic Press.)

	Methods[a]	MRPP test statistic	*p*-value
Spring Creek	1, 2, 3	−6.040	0.000478
	1, 2	−4.112	0.00712
	1, 3	−3.645	0.0111
	2,3	−5.682	0.00176
Poudre River	1, 2, 3	−14.762	1.06×10^{-8}
	1, 2	−17.875	1.93×10^{-8}
	1,3	−5.956	0.00143
	2,3	−8.734	9.16×10^{-5}

[a]Method: 1 = HPF; 2 = board; 3 = camera.

Table 6.3 Sample size and field time required to obtain an estimate within 10% of the mean of all data for respective methods. (Reproduced with permission from Dudley, S., Bonham, C., Abt, S., and Fischenich, J. 1998. Comparison of methods for measuring woody riparian vegetation density. *J. Arid Environ.* **38**: 77–86. © Academic Press.)

Site	Method	Ave. field time/measurement (min)	Sample size	Field time (min)	Rank
Spring	HPF	3.5	40	140	3
Creek	Board	1.5	15	23	1
	Camera	3.7	20	74	2
Poudre	HPF	3.2	25	80	3
River	Board	1.3	15	20	1
	Camera	3.5	15	53	2

simple randomized process of locating each individual pin minimizes this correlation and, thus, reduces interspatial effects of pins in frames and on line-point transects. To see this, consider that a cluster sampling design (arranging pins into frames or along lines) compared to a simple random sampling design (in which randomly located single pins are used) will give $S_w^2 < S_B^2$.

S_w^2 is the within-cluster variance and S_B^2 is the between-cluster variance. So, one should use simple random sampling to locate individual points. Often vegetation ecologists assume that this latter design is too costly in time units compared to gathering more data faster with a frame or line. In fact, use of a single point distributed over the extent of a plant community will provide more precise estimates of the species composition of the species in that community.

For a single pin as a sampling unit, one should consider the frame that Owensby (1973) described to minimize observer bias in single-point placement. Place a 3-legged, point frame at an individual location with one leg of the point frame placed

Table 6.4 Variances for cover of individual species sampled by 100 randomly located points compared to 100 points in 10-point frames. (Reproduced with kind permission from Goodall. 1952; Point quadrats; *Australian Journal of Scientific Research. Series B, Biological Sciences* **5**:11. © CSIRO.)

Species	10 Frames	100 Frames
Sedge (*Carex hebes*)	35.03[a]	10.92
Australia bluegrass (*Poa caespitosa*)	35.06	9.54
Purple violet (*Viola betonicifolia*)	16.57	16.48
Mountain woodruff (*Asperula gunnii*)	37.83	8.23
Yam-daisy (*Microseris scapigera*)	21.12	3.61

[a]Variances calculated after arc sine transformation of percent cover values.

at the end of the predetermined number of paces. The point frame is leaned backward on two of the legs toward the sampler and is leaned forward until point contact is made with a plant part, litter, or bare area. Three people are needed for efficient sampling. While these people can read 3000 to 4000 points in a day in practice, only 300 to 400 points are probably needed per plant community, as indicated by Goodall (1952) above.

Angle of placement

Most work on pin angles and placement has been to estimate vegetation cover with reference to the area of ground below plant canopies of grassland species. Generally, vegetation ecologists working in rangelands have paid little attention to the leaf area index (LAI) and its emphasis on "cover per unit of ground area." This index is important to describe photosynthetic characteristics by species as well as for nutrients needed for plant growth in a given area. Warren-Wilson's work (1963a) brought attention to properties of pins and their documented influences on results of pin angle degree and their diameters in an estimation of LAI (as reported under "Measurement of foliage pattern with point methods" below). Later, Groeneveld (1997) showed that pin angle and placement were not necessary requirements to obtain unbiased results. Instead, he used the vertical pin alignment in the point-frame with the standard 10 pins spaced 10 cm apart to obtain estimates of leaf area index (LAI) from 0.5×0.5 m plots. The point frame was superimposed in a two-dimensional fashion over each 10 cm increment of the plot edges for a total of 100 points. Both the frequency and the LAI were calculated from the data. His study showed that a "correction" factor could be applied to frequency values for species to give LAI estimates, for eight species growing in the Great Basin of the USA. Six shrub species and two grass species were used to test the "correction" factor on the LAI of widely divergent species. Groeneveld suggested that an approximation of LAI by species could be obtained by multiplying frequency of a species by 2 as follows:

$$\mathrm{LAI} = 2(F_i) \tag{6.9}$$

where F_i is the frequency value for species i obtained by point sampling with a point frame and pins are vertical to the ground surface. This estimation method for LAI may be sufficient for practical purposes such as relations of plant water use over large areas (Groeneveld *et al.* 1985). If plant ecologists prefer not to make assumptions as those made by Groeneveld (1997), then refer to Goel and Strebel (1984) for detailed information on the use of the two-parameter beta- distribution model to study the relationship of leaf angles and estimation of LAI for several plant species, including grass and crop species.

There may be other reasons for using vertical, horizontal, or inclined placements of pins because each does influence the value of cover estimates and LAI. Plant ecologists interested in ecophysiological aspects of species within a plant community may need to separate LAI by individual species with respect to measurements with regard to time of day, season, or other important factors that account for differences in the LAI of species. Vertically positioned pins tend to hit flat-bladed species, such as forbs, more often than grasses. Inclined pins tend to favor grasses (Winkworth 1955). Tinney *et al.* (1937) recommended that pins be inclined at a 45° angle for grasses, and Warren-Wilson (1963b) recommended that an angle of 32.5° be used, which minimized underestimates of leaf area for grasses. Leaf size for trees also will give different results and angles of inclination need to be determined for each forest type when a periscope or similar instrument is used.

Diameter of pins and points

It may be instructive to know how to determine effects of leaf characteristics on estimates of true foliage areas (LAI) under experimental conditions. Errors result from finite diameter points and the magnitude of the error depends upon the leaf size and shape. Errors resulting from the finite diameter of the points are greater for leaves with a high perimeter:area ratio (Warren-Wilson 1959b). This error is expressed as a percentage of the true foliage area as

$$E = \frac{100d}{lb}(d + l + b) \tag{6.10}$$

where E is the error, d is diameter of the point, l is the length of leaf, and b is the breadth of the leaf. Leaves are considered to have elliptical shapes or to be a combination of ellipses. Percentage errors estimated by Equation (6.10) and experimentally determined by comparing areas that used 2 mm pin diameters and true point diameters are similar. Experimental errors range from 13 to 340% and are highest for a narrow leaf species and lowest for a broadleaf species. Error depends more on leaf width than on length.

Error in percentage cover is determined only if an assumption is made about the way in which the foliage is dispersed. Assume that the foliage is randomly dispersed. Then

$$C = l - e^{-r} \tag{6.11}$$

where $C = 0.01 \times$ percentage cover and $r = 0.01 \times$ relative frequency. Use of a pin of a certain diameter increases the relative frequency to an extent that is found in Equation (6.10), provided b is known. Hence, the increased percentage cover resulting from increased point diameter can be found. Thus, Equation (6.10), with its assumption of elliptical leaf shape, does provide an adequate estimate of error. The error is doubled for each halving of leaf breadth or doubling of pin diameter for long, narrow leaves. The error is slightly more than double for round leaves.

Measurement of foliage pattern with point methods

Pin angle is important if one is interested in the spatial patterns of leaves and associated cover within canopies. Warren-Wilson (1959a) used both vertical and horizontal pins to analyze the spatial (or pattern) distribution of foliage. He noted that pin points underestimated cover for erect-leaf species and overestimated cover for species with nearly horizontal leaves. The angle of the leaves, relative to ground level, of most species varies with environmental conditions. Warren-Wilson (1959a) suggested that the angle between the leaf and the horizontal pin can be expressed as:

$$\tan\alpha = \frac{\pi}{2} \times \frac{F_{\mathrm{h}}}{F_{\mathrm{v}}} \tag{6.12}$$

for flat leaves at a fixed angle where F_{h} and F_{v} are the number of contacts with foliage per unit length of pin (contacts per centimeter) set for horizontal and vertical quadrats, respectively. Cover may be estimated as foliage denseness where foliage denseness F is the total area of foliage per unit volume of space and is estimated as

$$F = \sqrt{\frac{\pi^2}{4}\left(F_h^2 + F_v^2\right)} \tag{6.13}$$

Assumptions with a two-dimensional point are that leaves have no thickness, face all directions with equal frequencies, and all leaves slope at the same angle within a particular layer. However, errors introduced by violations of these assumptions are not serious. Therefore, F is useful in describing patchiness in foliar cover, while the angle of leaf orientation is useful in interpreting how the patches occur with respect to leaf angle. Then, the pin angle to obtain cover estimates can be selected for each type of species and combination.

Because leaf area index (LAI) is the area of foliage per unit area or volume of foliage per unit of ground, foliage denseness F (Equation 6.13) should be used in place of the traditional method of using angular foliage contacts by a pin or laser beam. The definition of F in terms of foliage volume per unit of space is probably the best measure to study water relations in plant growth and to predict precipitation runoff in watershed models.

6.3 Line-intercept methods

Development of the line-intercept method has been attributed to Tansley and Chipp (1926). The method consists of measuring the intercept of each plant under a line, and the line is usually placed on the ground (Figure 2.3). Cover of grasses and forbs is measured on the line at ground level. The intercept of tree species crowns or boles also can be measured with a tape and recorded by species or species groups. Crown spread for tall trees is difficult to measure precisely. This difficulty can be overcome to some extent by use of a sighting instrument to project the edge of the crown onto the tape by holding the instrument vertically from the tape to crown edge of the tree.

Lindsey (1955) described a "sighting level" for vertical projection of crown outline onto the tape. It consists of a 1.5 m long stick with a screw mounted into the top end and a carpenter's level mounted 30 cm from the lower end of the stick. The carpenter's level is used to control the vertical direction of the stick. The crown outline is a source of large variation. If the tree has a broken canopy within the intercept by the tape, gaps must be excluded from the measurement. Small gaps within the canopy can, however, be ignored.

Canfield (1941) described the line-intercept method to estimate cover of grassland and shrub vegetation. The sample unit was a line transect with length of 15 m for areas with cover of 5–15% and a 30 m line where cover is less than 5%. The intercept of the plants by species, through which a vertical plane must pass, was measured directly. The line was placed randomly to obtain the estimates and equipment included some type of line (wire, steel, tape, etc.) and two pins for securing the line tightly on either end. Total intercepts of the line by each species are calculated and expressed as a percentage of the total line length to give an estimate of percentage cover by species.

Accuracy and consistency in the use of the line-intercept method is highly influenced by the determination of the end points of the line. The longer the line, the more bias there is in data collection because the line cannot be stretched and maintained to an equal height above the ground. Even when there are additional support rods inserted at equal intervals along the line, there is still significant sag in the line between support rods. It is obvious that shorter length transects should be used as well as the calculation of sample adequacy needed to sample the plant community for cover.

The line transect can be of any length but must take into consideration the effects of length on species cover estimates because individual species are not distributed at random. Rather, each species forms an aggregation or clumped population that is fragmented into small units and usually creates several sizes of spatial patterns that are crossed by line transects. Transect length should not be increased as the vegetation cover decreases in an area. To do so would only increase the error of estimating cover because of the occurrence of varying pattern sizes among clumps of species. Furthermore, to select the transect length based on time needed to collect data along one transect is not an acceptable criterion. The denser the vegetation, the shorter the transect length should be, not the other way around, as suggested by USA rangeland researchers in the mid-twentieth century. No study has been found that reported on the effects of combinations of: (1) the time needed to place and read various lengths of transects, (2) number and lengths of transects placed and read within various densities of vegetation, and (3) size and number of spatial patterns of species and their combinations with other species along a transect of various lengths in varying vegetation densities.

The line-intercept method can only be applied to plants (trees, shrubs, grasses) that have a solid crown. Stephenson and Buell (1965) considered tree and shrub branches closer than 10 cm as solid crown. If the canopy is broken, cover of individual components should be summed and recorded as one entry. This facilitates the count of individuals for the purpose of density measurement. Therefore, the line intercept is more useful for estimating cover of shrubs and trees in open-grown, woody vegetation.

Line transect and point intercept

Point-intercept and line-transect methods are sometimes combined for estimating cover of plants in short vegetation types (Heady *et al.* 1959, Parker and Glendening 1942). Poissonet *et al.* (1972) used this method in tall grass vegetation. A line transect is placed, and the intercept is read at randomly selected points. Steel rods (pins), small in diameter, with sharpened end points are used at selected distances along the transect line to record an intercept. Selection of locations along the line transect to lower the pin is by a random process for estimation of species composition or a non-aligned systematic process (see Chapter 4). There is little difference in total number of contacts because only totals for transect length are usually of interest. Averaging over several transects by species' contacts provides a proportion of total hits by species (Table 6.5). Refer to Table 6.6 for sampling methods used to obtain data.

Bonham and Reich (2009) studied the influences of errors in total basal cover estimates that result from inexact relocation and orientation of a resampled point-line transect. Use of the point-line transect was twofold: (1) to obtain estimates of species basal cover and (2) to resample and determine changes, if any, in estimates of total cover values if the exact beginning point of the transect and the angle of orientation were not attained. A 2×2 m plot was located within a typical site of the short grass vegetation type. A total of 649 plants occurred within the plot (Figure 6.3) and each plant was measured for its x, y coordinates in the plot and its basal diameter. Four maximum errors in locating the starting point of a transect (0 cm, 0.5 cm, 1 cm, and 1.5 cm) were studied, as were 5 maximum deflection angles (θ) for the transect being placed on the original line ($\pm$) 0°, 0.5°, 1.0°, 1.5°, and 2.0°) (Figure 6.4). Fifteen point-line transects, each 1 m in length, were selected at random and permanently located along the baseline. Each of these permanent point-line transects was located

Table 6.5 Example of computations of percentage cover using the line-point sampling method using a 100 m transect and a point dropped at 10 cm intervals.

Species	Number of hits per line					Ave. hits	% cover	% comp.
	1	2	3	4	5			
Blue gamma	219	382	71	106	318	219	21.9	41.5
Western wheatgrass	43	17	33	81	27	40	4.0	7.6
Needle-and-thread	64	73	0	42	9	38	3.8	7.2
Buffalo grass	192	108	362	46	7	143	14.3	27.1
Sand dropseed	0	0	24	57	16	19	1.9	3.6
Sedge	76	62	12	119	42	62	6.2	11.7
Three awn	0	4	0	32	0	7	0.7	1.3
Total vegetation	594	646	502	483	419	528	52.8	100.0
Litter	187	132	272	231	318			
Bare ground	219	222	226	286	263			
Total	1000	1000	1000	1000	1000			

The data were recorded on a Loamy Plains Range Site, near Fort Collins, Colorado (species same as in Table 6.6).

Table 6.6 Example of computations of percentage cover by point sampling methods

Species	Number of ntercepts[a] in 1000 Point quadrats[a]	% Cover[b]	% Comp.[c]
Blue grama (Bouteloua gracilis)	237	3.7	38.4
Western wheatgrass (Agropyron smithii)	93	9.3	15.1
Needle-and-thread (Stipa comata)	42	4.2	6.8
Buffalo grass (Buchloe dactyloides)	128	12.8	20.7
Sand dropseed (Sporobolus cryptandrus)	27	2.7	4.4
Sedge (Carex spp.)	57	5.7	9.2
Three awn (Aristida spp.)	36	3.6	5.8
Total vegetation	620	62.0	100.0
Litter	138	13.8	
Bareground	242	24.2	
Total	1000	100.0	

Data collected near Fort Collins, Colorado.

[a]Sampling was by a point frame with 10 pins. Observations were recorded at 100 randomly selected locations. A pin lowered each time was considered a point quadrat. Data may also be analyzed by considering one point frame (10 pins) as one quadrat. This will give 100 random data points.

[b]Percentage cover = (no. of point intercepts/total points) × 100.

[c]Percentage composition = (no. of point intercepts/total point intercepts of vegetation) × 100.

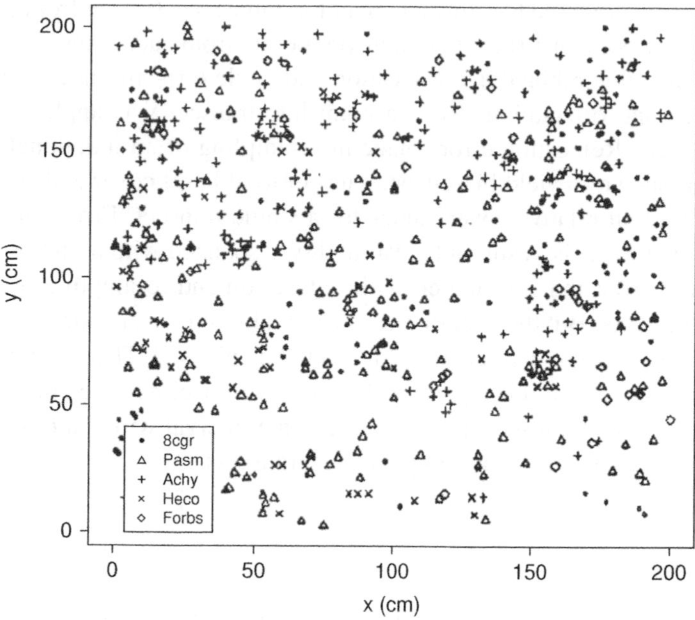

Figure 6.3 Spatial distribution of plants found within the 2 × 2 m plot by species. (Reproduced with permission from Influences of transect relocation errors on line-point estimates of plant cover. Bonham and Reich. 2009. *Plant Ecology* **204**: 173–178.)

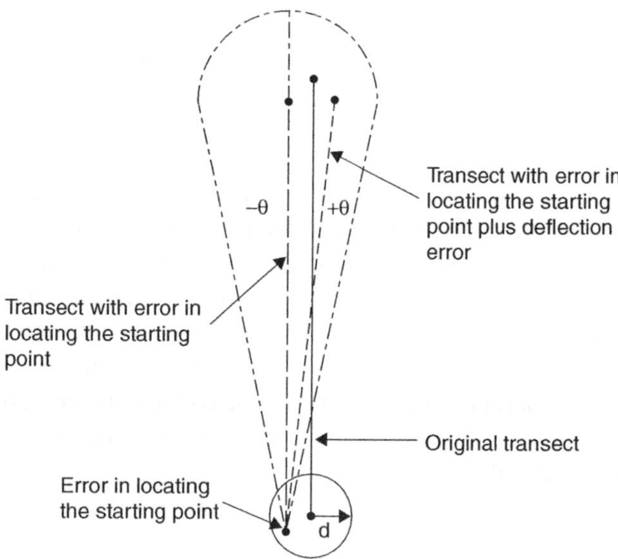

Figure 6.4 Graphic illustration of the simulation study of relocating and transect orientation. (Reproduced with permission from Influences of transect relocation errors on line-point estimates of plant cover. Bonham and Reich. 2009. *Plant Ecology* **204**: 173–178.)

perpendicularly along the baseline and read for contacts at 0.5 cm from the starting point and every point at 1.0 cm to obtain 100 points from each transect.

Each of the 19 combinations of location and angle error in relocating the starting point (d) and orientation (θ) of a point-line transect was applied to each of the 15 transects. Relocation errors made in resampling of an individual point-line transect was shown to yield biased estimates of total basal cover and species composition, based on relative cover values for a single transect. Errors in cover estimates obtained were the result of the variability associated with both the estimation of total plant cover and with the errors in relocation and resampling of individual point-line transects. But the overall means for total basal cover estimates from both resampled and original transects were shown to be unbiased. Therefore, it is recommended that estimates of change in total plant basal cover be based on an adequate number of point-line transects placed at each time interval within a monitored area rather than on permanently located point-line transects.

Laser pointer

A laser beam might be used in conjunction with a line transect to locate sample points. The beam is used as a pointer to cast a spot of light onto vegetation below. This method is thought by some investigators to be less biased and yield reproducible results in comparison to the use of a pin. Some users of the laser as a "pin point" erroneously believe that the height of the beam source does not have to be precisely located in the vertical plane each time, as does that of a pin. In fact, accuracy of either method is not affected by the beginning height of the beam or the pin, but rather by the failure to adhere to the exact location of the vertical plane as the observer moves from one location to another along a transect length.

Both point-line methods have the same problem with determination of a hit on vegetation below the upper layer of vegetation. It might not be possible to leave the lower layers of vegetation undisturbed as a search is made for additional contacts below the top layer of vegetation. There is a possibility that an observer can remove each part contacted, by clipping only the single leaf or stem contacted first by the pin or the light beam. Clipping has to be at the height level of the material contacted and this procedure continued for all subsequent contacts to ground level. Other problems with point sampling are common to any point contact, regardless of the method used to produce the point.

Line shadow transect estimation of cover

The line-intercept method to estimate shrub cover has been the standard method used by plant ecologists for over six decades. A canopy part intercepted along a

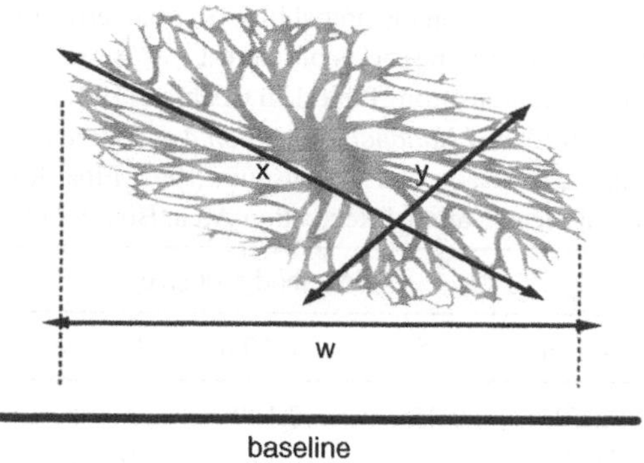

Figure 6.5 Canopy measurement method for line shadow transect data. (Reproduced from *Ecological Monitoring in Lake Mead National Recreation Area: Vegetation Change and Ecological Risk.* Bonham, C., W. Brady, and S. Bousquin. 1997. Final Report Volume I. The National Park Service, Lake Mead National Recreation Area, Boulder City, NV.)

line is used to estimate cover (Figure 2.3a). Thompson (1992) introduced a variation in the line-intercept method to estimate any measured variable. Brady (1997) and Brady and Eldridge (2008) used the label LST for the line shadow transect, to prevent confusion of these two line-intercept methods. Bonham *et al.* (1997) conducted a study to compare the line shadow transect method for shrub canopy cover estimates to estimates obtained from the traditional line-intercept method. In each case the canopy measurements used were from the same shrub. Six plots ranging in size 30 × 25 m to 60 × 25 m were replicated and located in white bursage (*Ambrosia dumosa*) communities in Arizona (USA). All individual plants of the shrub within these communities were measured for canopy lengths on the major and minor axes (an ellipse), as illustrated in Figure 6.5. Canopy areas were calculated from these variables.

The LST method was found to provide the most precise estimate of the shrub cover obtained from line-intercept transects and individual shrubs (same shrub used in all comparisons) in various plot sizes. This method (LST) consistently had the lowest standard errors over transects within sites for cover and the LST also had negligible bias effects on both type I and type II errors for comparing methods for cover estimates (Table 6.7).

Calculations for the LST method follow (see Figure 6.5 and Figure 6.6). All equations are from Thompson (1992) with shrub canopy area as the random variable. First, estimate the selection probability associated with each plant intersected along each transect

$$p_k = \frac{w_k}{B} \tag{6.14}$$

Table 6.7 Effect of bias in methods on the probability (p) of an error greater than 1.96 σ, where σ is the standard error. The shrub species of interest was Bursage (*Ambrosia dumosa*). Values are approximate interpolations from Table 1 in Cochran (1977). (Reproduced from *Ecological Monitoring in Lake Mead National Recreation Area: Vegetation Change and Ecological Risk*. Bonham, C., W. Brady, and S. Bousquin. 1997. Final Report Volume I. The National Park Service, Lake Mead National Recreation Area, Boulder City, NV.)

MP	Method plot size						
Bursage	LST	1 × 1 m	1 × 5 m	1 × 10 m	2 × 5 m	2 × 10 m	4 × 5 m
2.2	0.050	0.050	0.055	0.180	0.105	0.302	0.148
3.2	0.050	0.050	0.062	0.112	0.112	0.152	0.196
4.2	0.050	0.050	0.073	0.141	0.063	0.277	0.056
5.2	0.050	0.050	0.053	0.137	0.121	0.430	0.168
6.2	0.050	0.053	0.051	0.104	0.070	0.050	0.061

MP = Monitoring point; LST = Line shadow transect.

where $k = 1, 2, \ldots$ to the total number of plants intersected, w is the w length of the plant parallel to the baseline, and B is the baseline length (Figure 6.5 and Figure 6.6).

The area of each shrub canopy intercepted by the sample transect is estimated by

$$A_k = \pi(a_k x b_k) \tag{6.15}$$

where A_k is the area of a shrub canopy (k)/unit of ground covered by the crown of each intercepted shrub; a_k and b_k, respectively, are the major and minor axes of the

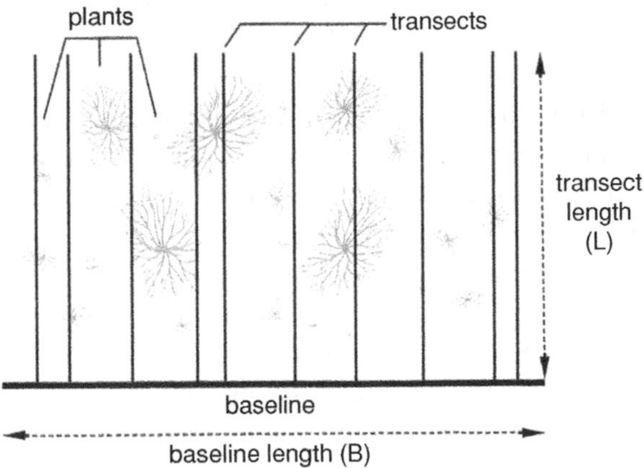

Figure 6.6 Field layout of the line shadow transect of shrub sampling. (Reproduced from *Ecological Monitoring in Lake Mead National Recreation Area: Vegetation Change and Ecological Risk*. Bonham, C., W. Brady, and S. Bousquin. 1997. Final Report Volume I. The National Park Service, Lake Mead National Recreation Area, Boulder City, NV.)

ellipse formed around the greatest canopy diameter and the next greatest canopy diameter that is perpendicular to the major axis (Figure 6.6).

The value A_k from Equation 6.15 is divided by the selection probability for that plant (Equation 6.14) to obtain the weighted area of the kth shrub canopy,

$$\lambda_k = \frac{A_k}{p_k} \tag{6.16}$$

where λ_k is the weighted shrub cover area, A_k is the area of the ellipse, and p_k equals the selection probability of the kth plant in the sample.

For each transect i in a sample of n transects, sum the weighted areas

$$v_i = \sum_{k=1}^{n} \lambda_i k \tag{6.17}$$

where v_i is the sum of weighted cover areas on the ith transect for the k plants which occurred on the transect. The total area of shrub cover A, for n transects/site, is the sum of $v_1, v_2, \ldots, v_n$ (Thompson 1992). Their sample mean is calculated as

$$\tau_p = \frac{1}{n}\sum_{i=1}^{n} v_i \tag{6.18}$$

where τ is the estimated total shrub cover/site with n transects. The sample variance of the v's is

$$s_v^2 = \frac{1}{n-1}\sum_{i=1}^{n}(v_i - \tau_p)^2 \tag{6.19}$$

and an unbiased estimate of the variance of τ is

$$\mathrm{var}(\tau) = \frac{s^2}{n} \tag{6.20}$$

An estimate of mean cover per unit area is then obtained from these estimators as

$$\lambda = \frac{\tau}{D} \tag{6.21}$$

where τ is the unbiased estimate of total cover, and D is the area of the study plot ($B \times L$). The variance of λ is

$$\mathrm{var}(\lambda) = \frac{\mathrm{var}(\tau)}{D^2} \tag{6.22}$$

6.4 Crown diameter and canopy closure method

Measurements of individual components of a tree are provided in detail by Husch *et al.* (2003). Tree canopy cover or canopy closure can be estimated from measurements of crown diameter. In many species, crown diameter is correlated with the trunk diameter at breast height (dbh). This relationship makes it easier to estimate crown diameter and, in turn, canopy closure from measurement of diameter at breast height (dbh) only. Directly related to volume, dbh is referred to as the *basal area* and is a good estimate of stand density (Husch *et al.* 2003). Thus, tree volume can also be estimated. In forest inventories, crown diameter is used as an independent variable in photo-stand volume tables. Dawkins (1963) collected data on 17 tropical tree species and examined the crown diameter:bole diameter relationships developed by other workers. He concluded that the most practical interpretation over the range of sizes for established trees was a straight line

$$y = a + b(x) \tag{6.23}$$

where y is crown diameter, x is the dbh, a is a constant, and b is the slope of the line. Where crown dimensions are measured directly, measurements are usually taken in at least two directions (longest and shortest dimensions of an ellipsoid).

Then area of crown or foliar cover is estimated as

$$\text{Foliar cover area} = \left(\frac{D_1 + D_2}{2}\right)\pi \tag{6.24}$$

where D_1 is the major axis length and D_2 is the length of the minor axis of the ellipsoid formed to circumscribe the tree canopy. Because cover is expressed as a fraction or percentage per unit area, unit area measurements are also taken simultaneously. Mueller-Dombois and Ellenberg (1974) consider this method impractical where a wide representative sample of cover by species is desirable over a large area. In these situations, use of the line-intercept method is more appropriate.

Tree crown diameters can be measured more effectively from aerial photographs compared to ground estimations (Husch *et al.* 2003). Such photographic measures appear smaller because gaps in a canopy are usually not distinguished. These authors suggested that such photos are a better measure of tree canopy because it is more highly correlated with tree and stand volume.

Photographs are not always available to the field ecologist, in which case, Johansson (1985) described a method to estimate canopy cover by vertical tubing. The tube consists of a 20 cm long hand-held brass tube of 1 cm diameter. It is mounted on a universal joint so that it hangs vertically. There is a cross-hair at the upper end of the tube and a mirror at the lower end. The cross-hair is sighted through the mirror and data are recorded for partial crown cover, full crown cover, or open sky. A "crown-free projection" (CFP) is calculated as

$$\mathrm{CFP} = \left(\frac{S}{T}\right) \times 100 \tag{6.25}$$

where S is number of open-sky sightings and T is total number of observations.

Other instruments that follow the general structural form described by Garrison (1949) are available, but this instrument is fairly easy to make and use. His device was called the "moosehorn," used to estimate crown closure (Figure 2.18). It is a type of periscope, and the top of the periscope has a grid of 25 dots on a glass plate. The periscope is held upright and fixed on a Jacob's staff. The instrument is completely leveled by a two-way level inside the periscope. The number of dots intercepted by a portion of the canopy is counted by viewing through a peephole. Note that 25 points provide an estimate of cover of ±4%. If higher precision is needed, then more dots must be used.

6.5 Variable plot methods

Vegetation measurement methods involving collection of data from a plot area delimited in two dimensions (i.e., length and breadth) are grouped together as plot methods. However, some sampling techniques for cover do not require a prescribed fixed area unit and are referred to as variable area sampling methods. Still, the measure of cover is related to an area of ground surface that can be estimated from a distance measure. For example, the methods below are founded upon the concept of an "area" included by the radius of an imaginary circle. The point and the line intercept transect are also, in a sense, variable plot sampling units. There are two other important variable area techniques used extensively for cover determination: (1) Bitterlich's "variable-radius method" for tree basal area and shrub canopy area, and (2) the "point-centered quarter method" (PCQ) to determine cover simultaneously with stem density.

Bitterdich's variable-radius method

This variable plot method was developed by Bitterlich (1948) in Germany to estimate canopy coverage. Grosenbaugh (1952), Hyder and Sneva (1960), Cooper (1957, 1963), and Fisser (1961) extended application of this method for use in tree, shrub, and grassland types of vegetation. The method was referred to as a "variable plot" because each plant counted represents a variable-sized plot that is proportional to the size of the plant. Trees or shrubs are counted in a circle from a central sampling point with an angle gauge (Figure 2.6). Plant canopies that are larger in diameter than the specified angle are included in the count, while others are ignored. The number of plants is proportional to their stem or basal area per unit ground area. As explained in Chapter 2, the angle gauge is held horizontally with the cross-piece facing toward the plants. All plants surrounding the sample point are targeted at a fixed height, usually breast height (1.4 m) and viewed through the other end of the gauge. Only those plants are counted for which thediameter exceeds the cross-piece. Bitterlich (1948) recommended a gauge ratio of 1.41:100 cm, and the tree count is divided by two to obtain an estimate of basal area in square meters per hectare. Grosenbaugh (Figure 2.6) recommended a still simpler ratio of 2:100 cm, in which case the tree count is also equal to basal area in square meters per hectare. In the past, this ratio has been used in plant ecological studies (Shanks 1954, Rice and Kelting 1955, Rice and Penfound 1959). Conversion of the ratio of basal area per area of ground sampled to percentage cover is given by

$$\%\ \text{basal cover} = \left(\frac{\text{Total basal area/unit ground area}}{\text{Total area sampled/hectare}}\right) \times 100 \qquad (6.26)$$

Point-centered quarter method (PCQ)

A number of distance methods have been developed for timber surveys to give density estimates (discussed in Chapter 7). However, in these methods, some additional data also can be collected to estimate basal area or canopy coverage by trees or foliage cover of shrubs. In the PCQ technique, four quarters are established at each sampling point. A cross of two lines, one in the direction of the compass and the other perpendicular to the compass line passing through the sample point, is established. Distance to midpoint of the nearest tree from the sample point and its dbh is measured in each quarter. The density estimate, as obtained from equations in Chapter 7, is multiplied by the average basal area to give the basal area per unit area. The method can be used to measure cover of shrubs and trees. In the case of shrubs, instead of dbh for trees, two measurements of a crown ellipse are recorded at right angles to each other, as in the crown ellipse method discussed earlier. Average foliar cover multiplied by density of shrubs gives total cover. The PCQ method has some limitations in its application to the measurement of cover for shrubs and trees. If small shrubs or trees are obscured by larger plants, counting may be difficult. Some errors also result from plants with irregular outlines of canopies. It has been demonstrated that with only half as many observations, the same precision level can be attained with the line-intercept method.

6.6 Semi-quantitative methods

Methods discussed thus far involve measurements to estimate cover. These methods involve little or no judgment on the part of the investigator and are, therefore, regarded as quantitative methods. The results obtained are generally reproducible by other investigators. Many plant ecologists, however, are interested in qualitative attributes of species rather than quantitative attributes. Some qualitative methods can be used for mathematical computations of data and, thus, are suitable for statistical analysis. These methods can appropriately be referred to as semi-quantitative. All historical cover measurements beginning in the late nineteenth century were originally obtained to describe plant communities, not to make statistical comparisons among these communities. As a result, early "measurements" of plant cover were either by the use of scales or intervals of values. Changes made in equipment have often resulted in close approximations to these early methods. A valid concern has remained over the decades about the precision of cover estimates obtained by any set of class intervals.

Peet *et al.* (1998) has conducted a review and interpretation of currently used methods to obtain visual plant cover estimates by class intervals. They found that an inherent characteristic of visual estimates of cover existed in how humans visually perceive cover on a geometric scale rather than on a linear scale. This is because our visualization is attuned to doubling and we can more easily discern the difference between values of 1% and 2% cover than we can for cover values between 31% and 32%. This observation implies that a plot area could be sub-divided into small units to determine more precisely visual estimates of vegetation cover. The recognition of how humans visually perceive cover led to examples where a relatively large plot area has been divided into 100 units of 1% each to obtain estimates to the nearest 1% (Bonham *et al.* 2004, Chen *et al.* 2006, 2008). Bonham *et al.* (2004) and Bonham and Clark (2005) divided a 50 × 100 cm into 1% unit areas of 5 × 10 cm (50 cm^2), while Chen *et al.* (2006 and 2008) arranged a 10 × 10 cm grid into 100 points, each of which represents 1% for contacts of points with plants beneath the transparent plastic plate. Refer to Section 6.2.2 "Grid-quadrat frame" for a discussion of data collection and analysis by Chen *et al.* (2006, 2008) and others.

Cover class methods

The range of cover values, 0–100%, is arbitrarily divided into a number of classes, each of which is assigned a rating or scale value. Of course, if a given class has a broad range of scalar values there is a greater consistency among observers in assigning a species cover value to the same class. Most class intervals are unequal and allow easier estimation of species cover to a unit of area (e.g., quadrat).

If a scale at the lower range in values is further broken down into shorter intervals, a more precise estimation of cover is made of less abundant species. The species occurring in the sample plot are assigned these ratings or scale values on the basis of the area occupied by species. When numbers of individual species also are obtained from sample plots, these values are referred to as a species diversity or richness index. The midpoints of each class can be used for statistical analysis of data, in which case interpretation is based on the assumption that actual data values are uniformly dispersed about these midpoints. This assumption is a strong one and, if violated, renders any statistical analysis useless.

In order to assign scale or rating values to each species, the observer makes a guess as to whether a species occupies more or less than 50% of the area of the sample plot. If it occupies less than 50%, the next question to ask is whether it occupies more than or less than 25%. Similarly, if it occupies more than 50%, it is determined whether it occupies more than or less than 75% of the area. Similar reasoning is continued until the range of the species cover value is narrowed down to a predetermined rating or scale. Even an experienced investigator might assign a species to a lower or a higher scale value than occupied by the species. Since the midpoints of each class interval are wide apart, there can be a large variation in data between investigators. Also, in a species-rich herbaceous community, errors in cover estimates are more likely with finer-scale intervals than with scales that have broad categories.

The use of ranges to place numerical data includes the assumption that the field ecologist can determine that a datum truly falls within a given range of values. Furthermore, it is assumed that the process is highly repeatable and reliable for other observers using the procedure. Any attempt to obtain a cover measurement of less than 1% is futile unless an interpretation of the ecological meaning for the species involved can be substantiated. It is clearly obvious that reporting any vegetation measurement data to the nearest decimal place is not meaningful. The following two methods for plant cover estimation continue to be among the most popular methods used in several countries of the world to describe plant communities during the twentieth century and into the first decade of the twenty-first century.

Braun-Blanquet class method

This scale gives a combined estimate of abundance and cover (Braun-Blanquet 1965). The sampling unit is called a "relevé" and its size is based on the minimal area concept. The first step is to familiarize oneself with the vegetation in the relevé and the second step is to prepare a species list, while the third, and last, step is to assign a value to each species from the cover-abundance scale. A species with less than 5% but more than 1% cover receives the value 1, and those with less than

1% receive a + (plus) sign. Rare species receive a notation of "r". The other areas of the plot are intervals of 5%, 15% and 25%.

Daubenmire class method

Daubenmire (1959) proposed a scale for estimating cover alone. A 0.1 m^2 plot (20 × 50 cm, inside dimensions) was found to be satisfactory. The method is also referred to as the "canopy coverage method." The frame (5 mm diameter steel) is painted so as to divide it into quarters, crosswise. In one corner of the frame, two sides of a 71 mm^2 area are painted. This allows reference areas equal to 5, 25, 50, 75, and 95% of the frame.

The method used by Daubenmire has been changed in various ways. Changes made in the frame and the cover scales of the frame include: (1) range in cover class intervals, (2) size of frame, (3) shape of frame, and (4) frame placements along transect lines. Cover class changes range from equal classes (e.g. 10% intervals) to those used by some of the USA federal land management agencies and, by others of a small 0.1% increment class at lower cover-scale values. Still other changes use unequal classes at both the lower and upper cover values and retain equal cover class ranges in between. Some of these changes essentially returned cover classes to the original cover-abundance scale of Braun-Blanquet.

Extension of a cover plot

Regulatory agencies in the USA responsible for successful mined-land reclamation require cover estimates to be made to the nearest 1%. This level of precision is also used in monitoring programs to determine effects of livestock grazing on grasslands and forested plant communities. This requirement led to development of smaller plots placed within larger plots to estimate cover at a 1% level (Bonham and Clark 2005). Because of the popularity of the original Daubenmire plot described above, it seemed reasonable to use this plot to obtain more precise visual measurements of cover. Making the 5% area of the 20 × 50 cm plot into a 1% area of a larger plot, 50 × 100 cm, accomplishes this.

A 50 × 100 cm plot contains five of a 20 × 50 cm frame, as shown in Figure 2.5 of Chapter 2. Then each plot of 20 × 50 cm placed within this larger frame will have proportional changes in scale associated with the original 20 × 50 cm frame. The result is a frame that includes five of the small frames. This combination was used to describe cover to the nearest 1% on reclaimed mined land in the southwestern United States (Bonham *et al.* 2004). An evaluation showed that repeatability could be obtained in cover estimation to the nearest 1% among observers. The method was described in statistical model form by Bonham and Clark (2005). The square shape of the smallest area of 5% was obtained to from a rectangular frame (5 × 10 cm) imbedded within the larger frame. This small frame forms an area of 1% of the total

Table 6.8 Example of cover data by species and categories using the extended Daubenmire method frame (from Figure 2.5). (Reproduced with permission from Plant cover estimates: a contiguous Daubenmire frame. Bonham, C. D., Mergen, and S. Montoya. 2004. *Rangelands* 26:17–22.)

	Sample Point: 1					
Cover type	Plot 1	Plot 2	Plot 3	Plot 4	Plot 5	Total
Total cover	18	12	17	14	19	80
Bare ground	2	8	3	6	1	20
Litter	7	4	8	6	10	35
Rock	0	1	2	0	1	4
Bocu	4	1	0	0	3	8
Scsc	2	0	3	5	1	11
Pasm	5	6	4	3	4	22
Total live plant	11	7	7	8	8	41

frame (50 × 100 cm) area. This area can be cut out of cardboard or thin metal and placed as a guide within each sub-plot to estimate species cover to the nearest 1%, as well as cover of litter and rock. Table 6.8 illustrates how the data is collected by sub-plot (20 × 50 cm) and summed over the five sub-plots by species to obtain cover value estimates for each species per 0.5 m^2 of area. It is suggested users of this frame consider (Bonham *et al.* 2004):

- This larger frame when subdivided as illustrated in Figure 2.5 should lead to increased repeatability for one or more observers and provide a reliable measure of cover to the nearest 1%.
- An observer should not use data values less than 1% for cover measurements just because of a seemingly large 1% area displayed within the individual plots.
- That data collected from the extended frame should not be used to compare results with other plot sizes and shapes or with other categories for cover. For example, to multiply the first sub-plot in the extended frame by a factor of five to convert all plot data into other sized frames will probably be biased.

Spatial sampling and bivariate correlation

Strenge (1998) studied large-scale spatial patterns in plant species and associated soils in a shortgrass site (130 ha). The site, 805 × 1610 m, was sub-divided into 98 cells of equal size, 115 × 115 m to create a non-aligned, systematic sampling pattern (Figure 6.7). One sample quadrat (11.5 × 11.5 m) was randomly located within each cell (Figure 6.7). Percentage cover was estimated for all plant species, litter, rock, and bare ground. Cover quadrats (20 × 50 cm) were placed at each of the nine sample points placed within each cell. For each cell, an average for cover

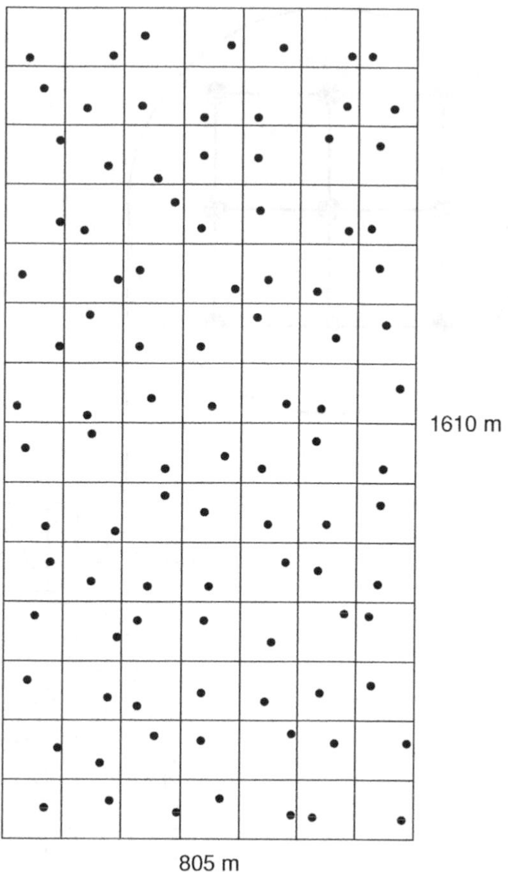

Figure 6.7 Sampling design arranged in a non-aligned systematic pattern. Each cell is of equal dimensions (115 × 115 m). Points within cells represent randomly selected points fom which data are collected. (Reproduced from Strenge, E. N. and M.S. Thesis. 1999. Colorado State University, *Fort Collins*. 134 pp courtesy of E. Strenge.)

values, along with soil variables obtained from composite samples collected from the center point, was used.

Moran's bivariate, I_{yx}, a cross-correlation index (Reich and Bonham 2001) for each paired combination of plant cover with each soil texture class, was calculated from

$$I_{xy} = \frac{\sum_{x=1}^{n}\sum_{y=1}^{n} w_{ij} y_i z_j}{W\left(m_w^2 m_z^2\right)^{1/2}} \tag{6.27}$$

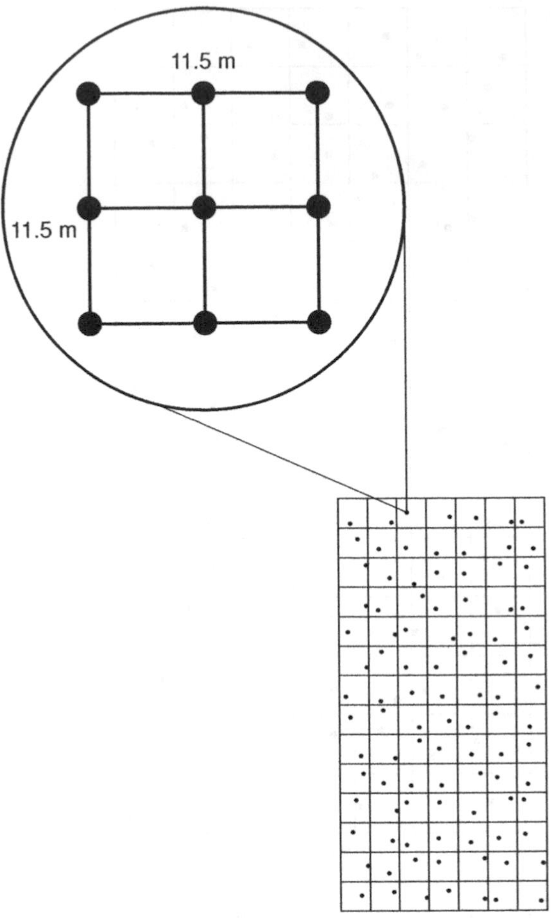

Figure 6.8 Arrangement of observations within sampling plots that were randomly located within each grid cell. Data on the percentage cover of plant species were collected at nine locations (black dots) and were then averaged for each plot. Data on plot biomass and soils were collected from the center point of each plot. (Reproduced from Strenge, E. N. and M.S. Thesis. 1999. Colorado State University, *Fort Collins*. 134 pp courtesy of E. Strenge.)

where w_{ij} is a scalar that quantifies the degree of spatial association between locations i and j, y_i is the value of variable y for plot i ($i = 1, 2, \ldots\ldots, n$) converted to a standardized value, z_j is the value of variable z for plot j ($j = 1, 2, \ldots\ldots n$) converted to a standardized value, W is the sum of all n_2 values of w_{ij}; m_y^2 is the sample variance of y_i, and m_x^2 is the sample variance of z_j. The value of I_{xy} is a weighted correlation coefficient between the variables y and z. If this value is significant and positive, then the two variables occur in similar spatial patterns. On the other hand, if there is a significant negative cross-correlation between the two variables, then the two variables probably do not occur in the same spatial pattern.

Table 6.9 Moran's spatial auto-correlation coefficients (I) and spatial cross-correlation coefficients (I_{xy}) for percentages of sand, silt, and clay within a native shortgrass steppe. Significant values of I_{xy} and I indicate the presence of spatial patterns, $p < 0.05$. (Reproduced from Spatial pattern analysis of vegetation and soil factors within the shortgrass steppe of North-central Colorado. Strenge, E. N., M.S. Thesis 1999. Colorado State University. Courtesy of E. Strenge.)

	% Sand	% Silt	% Clay
% Sand	$I = 0.05$	$I_{xy} = -0.04$	$I_{xy} = -0.04$
% Silt		$I = 0.06$	$I_{xy} = 0.02$
% Clay			$I = 0.04$

Cover values for two species are used to illustrate interpretation. Buffalo grass (*Buchloe dactyloides* (Nutt.) Engelm.) and red three-awn (*Aristida purpurea* Nutt.) were used with percentages of sand, silt, and clay contents of soils.

Significant auto-correlation was observed for each of the three soil particle size classes ($p < 0.05$) (Table 6.9). Silt and clay had similar spatial patterns, as indicated by their significant positive cross-correlation ($I_{xy} = 0.02$). On the other hand, sand ($I = 0.05$) had opposite patterns in its presence with silt ($I_{xy} = -0.04$) and clay ($I_{xy} = -0.04$).

Sand was positively spatially cross-correlated with red three-awn while buffalo grass was positively cross-correlated with both silt and clay (Table 6.10). The two species showed strong negative spatial cross-correlation with one another ($I = -0.12$.) (Table 6.11).

The overall study revealed that many of the variables measured were spatially auto-correlated, which indicated that spatial patterns existed for many plant species and soil characteristics. Moreover, cross-correlations existed between plant species and soil variables, both chemical and physical. Cross-correlations of

Table 6.10 Species that exhibited spatial cross-correlation (I_{xy}) with the percentages of sand, silt, and clay within a native shortgrass steppe. A significant I_{xy} indicates the presence of spatial patterns, $p < 0.05$. (Reproduced from Strenge, E. N., M.S. Thesis 1999. Colorado State University, *Fort Collins*. 134 pp courtesy of E. Strenge.)

Positive cross-correlations		Negative cross-correlations	
Sand	I_{xy}		I_{xy}
Red three-awn	0.07	Buffalo grass	−0.07
Silt	I_{xy}		I_{xy}
Buffalo grass	0.05	Red three-awn	−0.05
Clay	I_{xy}		I_{xy}
Buffalo grass	0.07	Red three-awn	−0.07

Table 6.11 Moran's spatial autocorrelation coefficients (I) for individual species and cross-correlation coefficients (I_{xy}) for each species and buffalo grass and red three-awn within a native shortgrass steppe. Significant I_{xy} and I indicate the presence of spatial patterns, at $p < 0.05$. (Reproduced from Strenge, E. N., M.S. Thesis. 1999 Colorado State University, *Fort Collins*. 134 pp courtesy of E. Strenge.)

Species	Moran's I	Buffalo grass (I_{xy})	Bed three-awn (I_{xy})
Buffalo grass	0.13	*	–0.12
Red three-awn	0.10	–0.12	*

species with each other and with soil properties indicated the existence of spatial patterns that closely resembled those of both plant and soil measures.

6.7 Bibliography

Bitterlich W. 1948. Die Winkelzahlprobe. *Allg. Forst-und-Holzwirtsch. Ztg.* **59**: 4–5.

Blackman G. 1935. A study of statistical methods of the distribution of species in grassland associations. *Ann. Bot.* **49**: 749–777.

Bonham C., Brady W., and Bousquin S. 1997. *Ecological Monitoring in Lake Mead National Recreation Area: Vegetation Change and Ecological Risk.* Final Report Volume I. The National Park Service: Lake Mead National Recreation Area, Boulder City, NV.

Bonham C. D., Mergen D. E., and Montoya S. 2004. Plant cover estimates: a contiguous Daubenmire frame. *Rangelands* **26**: 17–22.

Bonham C. and Clark D. 2005. Quantification of plant cover estimates. *Grassl. Sci.* **51**: 129–137.

Bonham C. and Reich R. 2009. Influences of transect relocation errors on line-point estimates of plant cover. *Plant Ecol.* **204**: 173–178.

Brady W. 1997. Line shadow transect (LST). In Bonham C., Brady W., and Bousquin S. *Ecological Monitoring in Lake Mead National Recreation Area: Vegetation Change and Ecological Risk.* Final Report Volume I. The National Park Service: Lake Mead National Recreation Area, Boulder City, NV.

Brady W. and Eldridge C. 2008. Adapting the line intercept to measure bunchgrass biomass. Presented at the 2008 Joint Meeting of the Society for Range Management and the America Forage and Grassland Council.

Braun-Blanquet J. 1965. *Plant Sociology: The Study of Plant Communities.* (Translated, revised and edited by Fuller C. D. and Conard H. S.) Hafner: London.

Canfield R. 1941. Application of the line interception method in sampling range vegetation. *J. Forestry* **39**: 388–394.

Chen J., Shiyomi M., Yamamura Y., Hori Y., and Yasuda T. 2006. Distribution model and spatial variation of cover in grassland vegetation. *Grassl. Sci.* **52**: 167–173.

Chen J., Shiyomi M., Yamamura Y., and Hori Y. 2007. Frequency distribution models for spatial patterns of vegetation abundance. *Ecol. Model.* **211**: 403–410.

Chen J., Shiyomi M., Bonham C., Yasuda T., Hori Y., and Yamamura Y. 2008. Plant cover estimation based on beta distribution. *Ecol. Res.* **23**: 813–819.

Cook C. W. and Bonham C. D. 1977. Techniques for vegetation measurements and analysis for a pre- and post-mining inventory. Range Science Department Series No. 28, Colorado State University: Fort Collins, CO.

Cooper C. 1957. The variable plot method for estimating shrub density. *J. Range Manage.* **10**: 111–115.

Cooper C. 1963. An evaluation of variable plot sampling in shrub and herbaceous vegetation. *Ecology* **44**: 565–569.

Daubenmire R. 1959. Canopy coverage method of vegetation analysis. *Northwest Sci.* **33**: 43–64.

Daubenmire R. 1968. *Plant Communities: A Textbook of Plant Synecology*. Harper & Row: New York.

Dawkins H. 1963. Crown diameters: Their relation to bole diameter in tropical forest trees. *Commonwealth Forestry Rev.* **42**: 318–333.

Dudley S., Bonham C., Abt S., and Fischenich J. 1998a. Comparison of methods for measuring woody riparian vegetation density. *J. Arid Environ.* **38**: 77–86.

Dudley S., Bonham C., Abt S., and Fischenich J. 1998b. Modification of the point frame for quantitative hydraulic investigations. *Rangelands* **20**: 25–27.

Evans R. and Love R. 1957. The step-point method of sampling: A practical tool in range research. *J. Range Manage.* **10**: 208–212.

Flippin-Dudley S. 1996. Vegetation measurements for estimating flow resistance. Ph.D. dissert. Colorado State University.

Fisser H. 1961. Variable plot, square foot plot, and visual estimate for shrub crown cover measurements. *J. Range Manage.* **14**: 202–207.

Garrison G. 1949. Uses and modifications for the "moosehorn" crown closure estimator. *J. Forestry* **47**: 733–735.

Goel N. S. and Strebel D. E. 1984. Simple beta distribution representation of leaf orientation in vegetation canopies. *Agron. Jour.* **76**: 800–802.

Goodall D. 1952. Some considerations in the use of point quadrats for the analysis of vegetation. *Aust. I. Sci. Res., Ser. B* **5**: 1–41.

Goodall D. 1953. Point-quadrat methods for the analysis of vegetation. *Aust. J. Bot.* **1**: 457–461.

Greig-Smith P. 1983. *Quantitative Plant Ecology*, 3rd edn. University of California Press: Berkeley.

Groeneveld D. P., Grate D. L., Hubbard P. J., Munk D. S., Novak P. J., Tillemans B., Warren D. C., and Yamashita I. 1985. A field assessment of above-and below-ground factors affecting phreatophyte transpiration in the Owens Valley, California. In *Proceedings of the first North American Riparian Conference*. USDA, Forest Service General Technical Report RM 120. 166–170.

Groeneveld D. 1997. Vertical point quadrat sampling and an extinction factor to calculate leaf area index. *J. Arid Environ.* **36**: 475–485.

Grosenbaugh L. 1952. Plotless timber estimates – new fast, easy. *J. Forestry* **50**: 32–37.

Hastings N. A. J. and Peacock J. B. 1975. *Statistical Distributions*. Butterworth: London.

Heady H., Gibbens R., and Powell R. 1959. A comparison of the charting, line intercept, and line point methods of sampling shrub types of vegetation. *J. Range Mange.* **12**: 180–188.

Huang D., Liu X., Song B., Shyomi M., Wang Y., Takahashi S., Hori Y., and Yamamura Y. 2007. Vegetation spatial heterogeneity of different soil regions in Inner Mongolia, China. *Tsinghua Sci. Tech.* **12**: 413–423.

Hughes G. and Madden L. 1993. Using the beta-binomial distribution to discrete aggregated patterns of disease incidence. *Phytopathology* **83**: 759–768.

Husch, B., Beers, T., and Kershaw, J. A. Jr. 2003. Forest Mensuration. John Wiley (and) Sons, Inc. : Hoboken, NJ.

Hyder D. and Sneva F. 1960. Bitterlich's plotless method for sampling basal ground cover of bunch grasses. *J. Range Manage.* **13**: 6–9.

Johansson T. 1985. Estimating canopy density by the vertical tube method. *Forest Ecol. Manage.* **11**: 139–144.

Kemp C. and Kemp A. 1956. The analysis of point quadrat data. *Aust. J. Bot.* **4**: 167–174.

Levy E. 1927. The grasslands of New Zealand. *N. Z. J. Agric.* **34**: 145–164.

Levy E. and Madden E. 1933. The point method for pasture analysis. *N. Z. I. Agric.* **46**: 267–279.

Lindsey A. 1955. Testing the line-strip method against full tallies in diverse forest types. *Ecology* **36**: 485–495.

MacArthur R. and Horn H. 1969. Foliage profule by vertical measurements. *Ecology* **50**: 802–804.

MacArthur R. and MacArthur J. 1961. On bird species diversity. *Ecology* **42**: 594–598.

Mielke P. W. Jr. 1986. Nonmetric statistical analysis: some metric alternatives. *J. Stat. Plant Inference* **30**: 377–387.

Mueller-Dombois D. and Ellenberg H. 1974. *Aims and Methods of Vegetation Ecology*. John Wiley & Sons, Inc.: New York.

Owensby C. 1973. Modified step-point system for botanical composition and basal cover estimates. *J. Range Manage.* **26**: 302–303.

Parker K. and Glendening G. E. 1942. *General Guide to Satisfactory Utilization of the Principal Southwestern Range Grasses*. Research Note 104, US

Department of Agriculture, Forest Service: Southwestern Forest Range Experiment Station, 4.

Pearse K. 1935. An area-list method of measuring range plant populations. *Ecology* **16**: 573–579.

Peet R., Wentworth T., and White P. 1998. Method for recording vegetation composition and structure. *Castanea* **63**: 262–274.

Poissonet P., Daget P., Poissonet J., and Long G. 1972. Rapid point survey by bayonet blade. *J. Range Manage.* **25**: 313.

Raunkiaer C. 1934. *The Life Forms of Plants and Statistical Plant Geography.* Oxford University Press: Oxford.

Reich R. M. and Bonham C. D. 2001. Spatial analysis of grazed white bursage in the Lake Mead National Recreational Area, Nevada, USA. *Grassland Sci.* **47**: 128–133.

Rice E. and Kelting R. 1955. The species area curve. *Ecology* **36**: 7–11.

Rice E. and Penfound W. 1959. The upland forest of Oklahoma. *Ecology* **40**: 593–608.

Shanks R. 1954. Plotless sampling trials in Appalachian forest types. *Ecology* **35**: 237–244.

Shiyomi M. and Yoshmura J. 2000. Measures of spatial heterogeneity for species occurrence or disease incidence with finite-counts. *Ecol. Res.* **15**: 13–20.

Shiyomi M., Gaborcik N., Koizumi H., Javorkova A., Uhliarova E., and Jezikova O. 2004. Spatial patterns and species diversity of plant communities in sown, oversown, and semi-natural grasslands in Banska Bystrica, Slovakia. *Grassl. Sci.* **50**: 1–8.

Stephenson S. and Buell M. 1965. The reproducibility of shrub cover sampling. *Ecology* **46**: 379–380.

Strenge E. N. 1999. Spatial pattern analysis of vegetation and soil factors within the shortgrass steppe of North-central Colorado. M.S. Thesis. Colorado State University, Fort Collins.

Tansley A. 1947. The early history of modern plant ecology in Britain. *J. Ecol.* **35**: 130–137.

Tansley A. and Chipp T. (eds) 1926. *Aims and Methods in Study of Vegetation.* British Empire Veg Chee and Crown Agents for the Colonies: London.

Thompson S. K. 1992. *Sampling*. John Wiley & Sons, Inc.: New York.

Tinney F., Aamodt O., and Ahlgren H. 1937. Preliminary report of a study on methods used in botanical analyses of pasture swards. *J. Am. Soc. Agron.* **29**: 835–840.

Warren-Wilson J. 1959a. Analysis of spatial distribution of foliage by two-dimensional point quadrats. *New Phytol.* **58**: 92–101.

Warren-Wilson J. 1959b. Analysis of the distribution of foliage area in grassland. In Ivins J. D., ed., *The Measurement of Grassland Productivity*. Butterworths: London, 51–61.

Warren-Wilson J. 1960. Inclined point quadrats. *New Phytol.* **59**: 1–8.

Warren-Wilson J. 1963a. Estimation of foliage denseness and foliage angle by inclined point quadrats. *Aust. I. Bot.* **11**: 95–105.

Warren-Wilson J. 1963b. Errors resulting from thickness of point quadrats. *Aust. J. Bot.* **11**:178–188.

Winkworth R. 1955. The use of point quadrats for the analysis of heathland. *Aust. J. Bot.* **3**: 68–81.

7 Density

"Density" in plant ecology is defined as the number of individuals of a given species that occurs within a given sample unit or study area. A simple count can be made of the number of individual plants with their canopy components arising from a common root stock. This is not to say that such a distinct individual is always obvious, but when this is not the case, the identified plant unit counted should be illustrated via a photograph or line drawing and included in the report of results. For example, if a density count of a sod-forming species is important, then a measure of density could be based on the total number of plant stems emanating from an indefinite common root stock that occurs within the sampling unit area.

Density is often used in a vegetation survey to describe a species' status in a plant community. Yet, there are several problems that could occur in obtaining an estimate of density. Included are the definition of an individual plant, as addressed above, size and shape of a sampling unit with associated boundary errors for inclusion of a plant within the plot area, and use of estimates from variable area plots.

7.1 Related measurements

7.1.1 Frequency

Another measure closely related to density is "frequency." The concept of frequency was developed and used by the Danish ecologist C. Raunkiaer (1909, 1912, 1934 papers). He referred to the number of times a species was encountered when walking through a stand (a homogeneous area of vegetation) (Stearns 1959). Frequency is defined as the ratio between the number of sampling units in which the species is present and the total number of sampling units (Chapter 5). This ratio is often expressed as a percentage. As the size of the sampling unit area (plot) is increased, the chance of a species occurring within that plot also is increased. Then, frequency is actually dependent on density and pattern of a species. The relationship between density and frequency is based on the proportion of absence of a species

Measurements for Terrestrial Vegetation, Second Edition. Charles D. Bonham.

(1 – presence) in an area (See Equation 5.4 in Chapter 5). Chapter 5 has more detailed discussion of this relationship.

7.1.2 Abundance

"Abundance" as a qualitative variable refers to an arbitrarily estimated range in numerical values which expresses plentifulness or scarcity of a species. These ranges are usually expressed by assigning the species to abundance classes. Braun-Blanquet (1932) listed five such abundance classes: (1) very sparse, (2) sparse, (3) not numerous, (4) numerous, and (5) very numerous. The qualitative measurement of abundance, then, is subject to personal bias. Some plant species, because of their form, color, economic importance, aesthetic appeal, or familiarity to the observer, may be overestimated. Reversal of some or all of these characteristics may result in an underestimate of how plentiful a species really is.

As a result of the foregoing, abundance should have a quantitative meaning. An appropriate definition, then, is the average number of individuals of a species per unit area for plots that contain the species. The definition is illustrated in Figure 7.1.

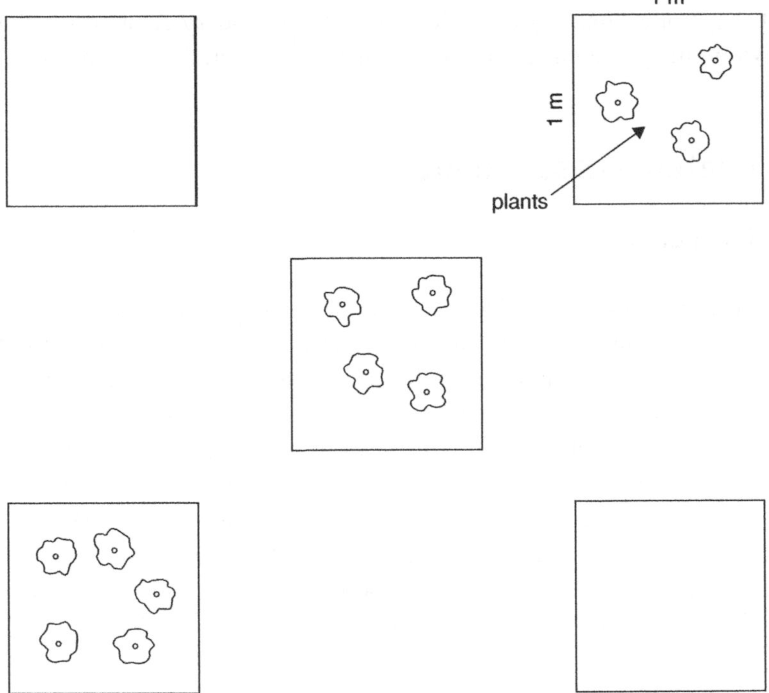

Figure 7.1 Abundance is equal to 12 divided by 3 ($A = 4.0$ plants/m^2).

In the example, abundance is 4.0/m^2 because only three quadrats contain the 12 individuals. An estimate of density is 2.4 individuals/m^2 (i.e., 12/5) while the frequency estimate is 3/5 or 60%. Thus, abundance also includes a combination of both frequency and density estimates.

7.2 Limitations of the density estimate

One must be able to recognize and define individual plants because this is critical to the measurement of plant density. Yet, identification of an individual plant is often one of the greatest difficulties in the determination of plant density. For instance, an individual plant may be defined as the aerial parts corresponding to a single root system (Strickler and Stearns 1962). This is easy for trees and other single-stemmed species such as annuals because their individual stems are readily determined. This may not always be true because annual and perennial grasses, when excavated, may show a wide range in variability in the number of culms for each root system. On the other hand, bunchgrasses may form easily recognized individual bunches.

An experienced student of plant ecology knows that old and large bunchgrasses break into smaller individual groups of culms and make the determination of individual plants difficult (Walker 1970). Moreover, perennial grasses and forbs sometimes spread laterally on ground surface (i.e., stoloniferous or rhizomatous forms) and are not possible to recognize as individuals. On the other hand, shrubs can be close together and produce both single- and multiple-stemmed plants that make density estimation difficult. If the definition of the individual is difficult, then the density measure should not be used to describe vegetation characteristics of an area because a reproducible measure cannot be obtained.

Again, the only practical definition of the counting unit for plants is the individual, whether it is a stalk, a culm, or a bunch of culms. The value of defining individuals depends on the purpose of the study, the definition of the unit, and the precision of the count. Use of a stem or shoot, as an individual, has the advantage of less variation in unit size than is the case with the use of either bunch or rhizomatous plants. Furthermore, the number of stems or culms may have a higher correlation with other measurements, such as biomass amount or basal cover.

A density estimate alone is a limited measure to determine plant community dominance of a species. This is obvious because density is an estimate only of the number of individuals per sample unit area and does not indicate the relative contribution of each species to total ground cover or total plant biomass produced in the vegetation type. Species contribution to total biomass may vary significantly from year to year as a result of differing environmental conditions. Yet, density may not change much. Annual species vary in number of plants, especially annual herbs, according to seasonal precipitation and temperature. On the other hand, numbers of some perennial plant species tend to remain fairly constant over time. This is true for shrubs, trees, and some perennial grasses and forbs. In spite of density estimate limitations, counting remains one of the easiest quantitative species measures to understand. Estimates of density are useful for monitoring plant responses to various vegetation treatments such as defoliation or environmental perturbations.

Three major techniques are used to estimate density: plot (quadrat), distance measures, and line transect. Plots and distance techniques are the most often used to estimate species densities. The following sections represent a fairly complete presentation of density estimation techniques. Some of the distance measures and subsequent estimation of plant density may be somewhat difficult to understand for the reader who does not have a statistical background. Yet, inclusion of this material is provided for techniques that result in unbiased estimates of density, especially for certain individual tree and shrub species.

7.3 Quadrat techniques

The number of individuals found within quadrats or plots can be counted to estimate density. Pound and Clements (1898) were among the first to use quadrats to measure vegetation characteristics such as density. They used 5 m^2 frames to count plant species in some vegetation types, but they mistakenly thought that estimation of density is affected by the degree of dispersion of individuals and quadrat size and shape. To assist interpretation of density estimates, two sampling characteristics for measurements should be considered: (1) distribution of the plants within the area to be sampled, and (2) the number of observations needed to obtain an adequate estimate of density. This latter characteristic eliminates the need to consider size and shape of the sampling unit.

7.3.1 Distribution

A measure of the distribution of individual plants of a species is important for interpretation of a density estimate. In populations of non-randomly distributed individuals, density is difficult to evaluate by simple techniques. The major source of difficulty occurs when plants are assumed to be randomly distributed and follow a Poisson distribution. This assumption is seldom satisfied because plants are aggregated into clusters of varying scales of patterns across the landscape. In which case, most sampling problems usually can be resolved by increasing the number of quadrats (sample size) used for counting density. A 1 m^2 plot probably would include most of the small-scale patterns of ecological information needed to interpret density for herbaceous plant species. A 10×10 m^2 plot should suffice for shrubs and trees in most environments. These plots should be arranged in a sampling design to cover the plant community. Then sample adequacy on a statistical basis should be estimated after density is obtained from 10–15 plots. Quite often, this preliminary sample size will suffice as estimated from the sample adequacy, Equation (3.43) in Chapter 3. If not, then the field ecologist should update a sample size estimate with the most recently obtained estimates of mean and variance and continue this iterative process until adequacy is attained.

As stated above, randomly distributed plants can be described by the Poisson series which is a rare-events mathematical series. It follows that a species is considered rare if it seldom occurs in a quadrat, then precision of the density estimate depends only on the number of individuals counted. The following equations illustrate these concepts (Greig-Smith 1983). Let x be the number of individuals that are counted in n quadrats. Then

$$Density = \frac{x}{n} \tag{7.1}$$

The variance s^2 is equal to its mean in a Poisson distribution. Then

$$s^2 = \frac{x}{n} \tag{7.2}$$

and the variance of the mean of n quadrats is

$$s_{\bar{x}}^2 = \frac{1}{n} \left(\frac{x}{n}\right) = \frac{x}{n^2} \tag{7.3}$$

The standard error of the mean is

$$s_{\bar{x}} = \frac{\sqrt{x}}{n} \tag{7.4}$$

The ratio of standard error of mean (Equation 7.5) to the mean (Equation 7.2)

$$\left(\frac{\sqrt{x}}{n}\right)\left(\frac{n}{x}\right) = \frac{1}{\sqrt{x}} \tag{7.5}$$

It is obvious that the standard error of the density estimate remains the same for the same number of individuals counted. This is true regardless of the size or number of quadrats used. Since a single observation from a randomly distributed population can be regarded as an unbiased estimate of the population parameter, the sample variance and standard error can be estimated from the number of individuals that occurred in the single quadrat. Let x be the number of individuals in a single quadrat as before. Then, x and $\sqrt{x}$ are estimates of the variance and the standard error, respectively. Note, however, that this single observation may deviate significantly from the true mean and therefore should be regarded as only a rough approximation of the population mean. In most natural populations, individual plants may not be randomly located, but occur in clumps. In fact, random populations of plants are the exceptions in nature, but may be found in individual species in many or even most plant communities, as indicated from spatial statistical analysis (Bonham *et al.* 1995, Strenge 1998).

Quadrat size affects the data distribution (whether Poisson, binomial, etc.), not the field distribution of plants. For example, a large quadrat size will result in a more symmetrical data distribution that can be approximated by the normal distribution and reduce the magnitude of the variance in comparison to a small quadrat. Intuitively, it is seen that more individual plants may occur in a large quadrat compared to a small quadrat, which allows for the occurrence of few individuals. In turn, the use of small plots could provide "rare-events" for data distribution described by a Poisson distribution.

7.3.2 Quadrat size and shape

Finding an adequate quadrat size for a plant species population of density values can be difficult, in which case one should consider several factors before selecting a given plot size and shape. Some of these factors are presented in Chapter 2. *First*, an important point to recall is that a relatively small plot has greater length of boundary per unit area than successively larger ones. One result of this relationship is that as the plot size becomes smaller, the greater the opportunity for boundary error. That is, an observer may consistently include individual plants near or on the boundary that should not have been counted, or might not count those that were within, but near the boundary. For this reason, a quadrat should not be small, say 0.1 m^2, if density estimates are made. *Second*, the shape of the data distribution curve (see Chapter 3) affects significance tests which are based on a normal distribution. Furthermore, plot size affects the data frequency curve shape, and as a result, efforts should be made to use an appropriate plot size, compared to a small plot area, that will decrease the variance for numbers of plants between observations. As the data values narrow in range within a plot, variance will decrease and density data distribution will approximate a normal distribution curve. *Third*, field experience will indicate, even to the novice, that counting large numbers of individual plants can be tedious and lead to inaccurate counts in large plots. A large quadrat can be subdivided or individuals marked as they are counted, but the purpose of using a large quadrat is then defeated. Sub-division of such quadrats only leads to increasing boundary lengths per unit area of new sub-quadrats. This problem can be solved by increasing the number (sample size) of quadrats used for counting.

Bartlett (1948) found that the most efficient size of quadrats for estimation of density corresponds to about 20% absence. This value for absence is the same as the product of quadrat size and plant density of about 1.6. This constant, 1.6, is obtained from the relationship between density and frequency of a species found in a random distribution. If

$$\text{Density} = -\ln\left(1 - \frac{F}{100}\right) \tag{7.6}$$

where F is the percentage frequency, and ln is the natural logarithm, then 1.0 – 0.80 = 0.20, which is the absence of the species. Then

$$\text{Density} = -\ln(1.0 - 0.80) = -\ln(0.20) = 1.61$$

Bartlett suggested that the efficiency is maintained over a size range of area, 0.7 – 3.0 m^2. This method of determining quadrat size is applicable when the distribution of plants is random or nearly so and may be used when the relationship between the logarithm of percentage absence and plant density is linear, which in practice it is not.

Quadrat size and shape is known to affect the precision of a sample estimate of density. Many ecologists have observed that more species are included in long, narrow quadrats because of the tendency for vegetation to be aggregated. However, aggregation of plants is two-dimensional, not one. In fact, optimum quadrat size and shape depends on the distribution of the species measured, and in general, larger quadrats are recommended for use in sparse vegetation. Although small plots are generally more efficient statistically, they often yield skewed data for density. The greatest sampling efficiency is generally indicated by the plot that provides the smallest variance for any specific measurement.

Eddleman *et al.* (1964) tested plot sizes and shapes to estimate density and frequency in dry-mesic alpine vegetation in northern Colorado. Plots were 100, 400, 800, and 1600 cm^2 with rectangular, square, and circular shapes. Two estimates of efficiency were made. One used the time required to sample for a 10% standard error of the mean. The other was the product of the average weighted standard deviation for a particular plot shape and size multiplied by the time required to read sufficient plots within 10% of the mean density (Table 7.1).

Plots that required the fewest number were a 400 cm^2 circle, 40×40 cm, and a 10×40 cm rectangle. Rectangular quadrats of the same area as square plots needed fewer observations in two out of three cases. Rectangular quadrats were more efficient for some species while square ones were more efficient for others. Standard deviations increased with increasing quadrat size, but when adjusted for quadrat size, larger quadrats had smaller standard deviations. Time to sample was less for medium to large plots and for rectangular plots. The 400 cm^2 plots of the Eddleman *et al.* study are favored because there is less likelihood of miscounting species.

Table 7.1 Efficiency estimates and ranking of quadrats. (Republished with permission of Torrey Botanical Society Inc., from An evaluation of plot method for alpine vegetation. Eddleman *et al.* 1964. *Bull Torrey Bot Club* 91: 446; permission conveyed through Copyright Clearance Center, Inc.)

Plot[a]	Ave. no. of plots for density stability points	Time (min) to read plots for density stability points	Average weighted standard deviation
10 × 40 cm R	51	69	5.08
20 × 80 cm R	56	120	3.44
40 × 40 cm R	48	100	3.47
400 cm^2 C, R	40	54	5.93
20 × 20 R	63	85	6.68
20 × 40 R	73	131	4.63

[a]C = circle; R = rectangle.

Evans (1952) discussed the influence of quadrat size on the distributional patterns of plant populations. Quadrats of sizes (m^2): 16, 8, 4, 2, 1, 0.5, 0.25, 0.125, and 0.0625 m^2 were used. The species studied had aggregated populations, and density estimates on an area basis were found to be the same for all quadrats.

7.3.3 Strip or rectangular quadrats

A strip quadrat is a rectangular-shaped quadrat that has one long dimension, length, which is exaggerated relative to its width. Some literature references make use of the term "transect" to describe the sampling unit. References also are made to uses of a "belt transect." Technically, all line transects have width and, in reality, are long, narrow quadrats, or equivalently, strip quadrats. These long quadrats can be subdivided into smaller units that can then be used to study the relationship between quadrat size and density estimates. The width of the strip might vary from 0.5 m wide in grasslands to 10 m in forested communities. The combination of widths and lengths to form quadrat areas will ultimately depend on plant diameter and spacing. Tree densities are often estimated by use of strip quadrats and are enumerated within diameter size classes. These size classes for diameters are used by foresters to estimate timber volume ready for harvest. In plant ecology, density plots for herbaceous plants can be from 0.10 m^2 to 1 m^2, while for trees and shrubs plot sizes for density range from 16 m^2 to 100 m^2 or greater.

Any quadrat of small size can be used as a modification of the strip-quadrat method. Simply place small quadrats at predetermined intervals along a transect. The advantage of using a transect is to ensure a sample from the entire stratum or vegetation type. The size of the quadrat, which is a segment from a strip quadrat, depends on the size and spacing of trees or shrubs. Statistical analysis is then conducted as if the strip quadrat had a mean density estimated from the mean of the individual small quadrats placed along transects. The estimate of density variance within an individual strip quadrat might be biased unless the small units are randomly placed along the transect or a spatial auto-correlation model is used in data analysis to account for spatial effects of plots being adjacent. However, no data analysis problem exists if interest is centered on estimates obtained from a larger (strip) quadrat; i.e., small plots within the larger strip plot can be averaged and this value used for the mean and variance estimates for the strip.

7.3.4 Considerations

Some users of the first edition of the present book (*Measurements for Terrestrial Vegetation*, 1989) pointed out that many of the references on specific measurements were taken from old literature. This is true because much the early work was conducted in northwest Europe, the United States, and Australia beginning in ca. 1905. Density estimates for herbaceous vegetation have been obtained frequently by a 1×1 m quadrat. Dice (1948) used a 0.8 m^2 quadrat to determine density in a grass-herb field. Smaller units, such as 20×50 cm, may be used in dense vegetation, such as meadows, annual grasslands, alpine-arctic, tundra, and sub-tropical vegetation types, where individuals are small and are in close proximity to one another. Hanson (1934) recommended the use of 0.1 m^2 plots for sampling a mixed prairie type, while Heady (1958) used 6.45 cm^2 plots in a California annual type. Eddleman (1962) and Bonham and Ward (1970) used a 10×40 cm plot to measure plant density for alpine-tundra vegetation.

Density estimates from forested areas are usually expressed as number of trees per acre or hectare. Lang *et al.* (1971) found 10×20 m quadrats for a tropical forest had the lowest required sample size for the majority of species counted. On the other hand, Bormann (1953) recommended the use of 10×140 m plots to determine density in hardwood forests, while Bourdeau (1953) used 10×10 m plots to determine density in a deciduous forest. These results are in contrast to those of Cain (1936), who used 50 m^2 plots for studies in deciduous forests. Mueller-Dombois and Ellenberg (1974) suggested the use of 10×10 m quadrats to determine tree density, and they further suggested a 4×4 m quadrat be used for sampling of woody undergrowth up to 3 m in height. The size of plot should always be sufficient to overlap small scales of patterns, as indicated by patchiness in vegetation.

7.4 Distance methods

Distance measurement techniques are used to estimate plant density (numbers of individuals/unit area) and have been used more extensively in forestry than in other vegetation type. These techniques are often referred to as "variable-plot" or "plotless" methods. There are no visible boundaries and measures depend on distances between points and between two plants or a combination of the two. The distance measure is usually considered as an estimate of the radius of an imaginary circle. Use of one or more of the distance techniques might save considerable time and could even improve the precision of the density estimate.

Density for plant species may be estimated by dividing the unit reference area (such as a hectare = 10 000 m^2) by the estimated mean area of an individual plant. Mean area, MA, is estimated by

$$\text{MA} = \frac{1}{\text{Density}} \tag{7.7}$$

As seen in Figure 7.2, the distance d extends the distance to the point or next closest plant. Thus, the distance measure (radius) is used to estimate area, which, in turn, is used to estimate density. Mean area is then defined as the reciprocal of density. In effect, an estimate of density is obtained from an estimate of mean area for an individual species. The remaining problem is to find a distance measure that will give the best estimate of the mean area per plant. In practice, the mean area is determined by averaging numerous distance measures for a species found in a vegetation type.

Techniques to obtain distance measures are grouped into one of two categories: (1) those to be used only for species that are randomly distributed (techniques include closest individual, nearest neighbor, random pairs, and the point-centered quarter) and (2) those used for species that are either randomly or non-randomly distributed. Some of the distance measures for density estimates can be used in

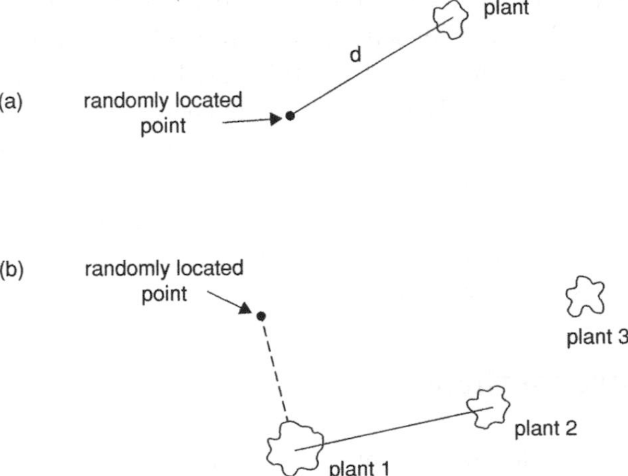

Figure 7.2 (a) Closest individual, point-to-plant method to obtain distance measure d; (b) nearest-neighbor method for measurement of plant-to-plant distance d.

both types of dispersion and include the angle-order method, the wandering-quarter method, and the corrected-point-distance method. Techniques to obtain distance measures from assumed random populations are provided in this chapter only if they provide unbiased estimates. Otherwise, emphasis is placed on methods that are used for both random and non-random distributions. The emphasis here is on the estimate being unbiased, not on the method of obtaining distance measures, as seen from Pollard's (1971) equations below.

7.4.1 Random populations

Individual plants and/or points should be selected in some form of randomness for measurement of distance. A purely random selection of plants involves the numbering of all plants of a species in a vegetation type and randomly selecting a sub-set of these numbers. Numbers in the sub-set represent plants from which a sample distance is measured to their nearest neighbor of the same species. If only total plant density is needed, then any species that is the nearest neighbor is measured for distance. It is obvious that pure randomization is too time-consuming to be of practical use. Therefore, there are several ways of selecting a random starting point from which measurements are made. Use some process that is free of personal bias to locate points in the field. It is important to note that consequences do exist if the spatial pattern of individual plants deviates considerably from that of a random population (Persson 1971). Namely, most of the estimators for density are seriously biased. If random distributions are assumed for plants, then the following techniques can be used to obtain various distance measures. Large-scale studies of plant species have found that dominant or co-dominant species occur as randomly distributed (Bonham *et al.* 1995, Reich and Bonham 2001). In practice, random distributions of individuals are assumed. This assumption does not always hold and should be checked quantitatively. In any case, the proper equation must be used, as mentioned above.

Closest individual method

Pollard (1971) presented an equation that provides for an unbiased estimate of density for data obtained from this method for distance measures. Note that distances are treated as radii of circles (r) and multiplied by pi (π) to obtain area. The closest plant to a random point method to estimate density then provides an estimate from

$$m = \frac{n-1}{\pi \sum_{j=1}^{n} r_j^2} \tag{7.8}$$

where n is the number of randomly located points and r represents individual distances to the nearest plant from the random point.

The variance of the estimated density is

$$\text{var}(m) = \frac{\lambda^2}{n-2} \tag{7.9}$$

and

$$\lambda = \frac{n}{\pi \sum_{j=1}^{n} r_j^2} \tag{7.10}$$

To estimate density by measuring the second or third nearest individual from a random point, the following equation from Pollard (1971) is used

$$m = \frac{K-1}{\pi \sum_{j=1}^{n} r_j^2} \tag{7.11}$$

where

$$K = \sum_{j=1}^{n} k_j \tag{7.12}$$

and k is the order of the individual measured. That is, $k = 3$ if the third nearest plant is measured from a point. Then, if $k = 3$ at every point and 50 points are sampled, $K = 150$. The variance of m in Equation 7.11 is

$$\text{var}(m) = \frac{\lambda^2}{K-2} \tag{7.13}$$

where

$$\lambda = \frac{K}{\pi \sum_{j=1}^{n} r_j^2} \tag{7.14}$$

It is more difficult to measure the third nearest plant to a random point compared to selecting the nearest plant to a point. There is no doubt that more opportunity for error exists in the determinations of the third and higher-order nearest plant. Then, from the practical viewpoint, the third nearest individual plant may be the furthest one to measure from a point, if indeed it can be identified as the third one in distance. The greater the density, the more difficult it becomes to determine the order in the distance of the plant.

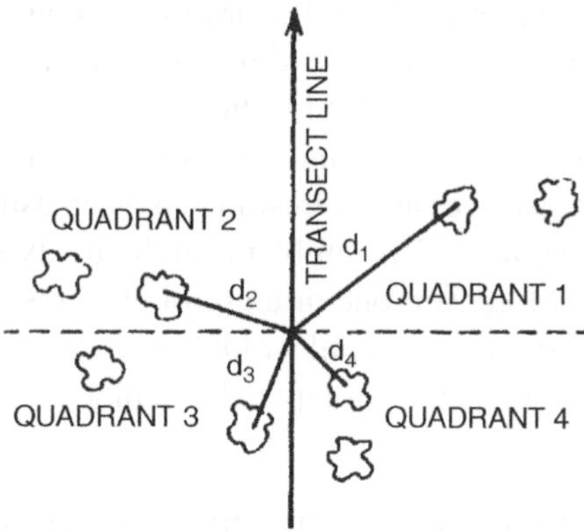

Figure 7.3 Point-centered-quarter method to measure distance d of plants in four quadrants from a point.

Point-centered quarter method

The point-centered-quarter (PCQ) method involves distances that are measured from a point to the nearest plant in each of four 90° sectors around a randomly or systematically established sampling point (Figure 7.3). Previously, the mean area occupied by a plant was determined by averaging the four distances of a number of observation points. Density was then determined by squaring the reciprocal of the average mean distance d per point. Note that no special attention is given to individual sample points with four measures of distance. One is made from each quadrant. However, density $= 1/d^2$ does not provide an unbiased estimate of density, while Equation 7.15 below does. Pollard (1971) suggested that an unbiased estimate for density is obtained from the PCQ method from

$$m = \frac{4(4n-1)}{\pi \sum_{i=1}^{n} \sum_{j=1}^{4} r_{ij}^2} \tag{7.15}$$

where n is the number of randomly located points and r is the individual distances to the nearest tree in a given quadrant. The variance of m is given by

$$\text{var}(m) = \frac{\lambda^2}{4n-2} \tag{7.16}$$

where

$$\lambda = \frac{16n}{\pi \sum_{i=1}^{n} \sum_{j=1}^{4} r_{ij}^2} \tag{7.17}$$

Pollard found that the PCQ method apparently has no advantage over the simple distance method (i.e., the distance to a plant from a randomly located point). This method was discussed in Seection 7.4.1. Pollard's equations solved problems associated with the other equations used to estimate density, but other problems arise from the PCQ method. For example, the locations for PCQ random points on steep terrain that have a mosaic of communities are difficult. Then density is usually underestimated because distances are overestimated when using a measurement and the tape is not perfectly straight. This problem can be partly solved by the use of a laser beam transit level. Moreover, one of a group of stems is recorded for plants. Most of these individual stems would fall within the imaginary plot and result in an underestimation by the PCQ method compared to a plot.

7.4.2 Distance methods for random and non-random populations

This group of distance measures, which might be used for randomly and non-randomly dispersed populations, includes the order method, the angle-order method (Morisita 1957) and the wandering-quarter method (Catana 1963). The angle-order method is not to be confused with the angle method presented above.

Order Method

Morisita (1957) described an order method that requires measurement to the *n*th nearest individual from a random point of origin. Table 7.2 contains the moments (mean and variance, μ and σ^2, respectively) from which density m can be estimated. Higher precision for an estimate is obtained by measurements to farther individuals since $(\sigma/r)^2$ decreases from the first closest individual to the second, third, and so on. A density mean estimate is obtained from

$$m = \frac{nN - 1}{\sum_{i=1}^{N} r_i^2} \tag{7.18}$$

Table 7.2 The density functions of distribution, the moments, and the values of $\frac{\sigma}{\bar{r}^2}$ in the order method. (Reproduced with permission from Morisita, M. 1954; Estimation of population density by spacing method. *Kyushu University Faculty Science Memorial Series E* **1**: 195.)

Order	Density function f(r)	$\mu' (= \bar{r})$	μ'_2	σ^2	$(\sigma/\bar{r})^2$
1st closest individual	$2m\,r\,e^{-mr^2}$	$1/2\sqrt{\pi/m}$	$\frac{1}{m}$	$0.21460\,\frac{1}{m}$	$0.8584\frac{1}{\pi}$
2nd closest individual	$2m^2\,r^3\,e^{-mr^2}$	$3\,(1/2)^2\,\sqrt{\pi/m}$	$\frac{2}{m}$	$0.23285\frac{1}{m}$	$0.4139\,\frac{1}{\pi}$
3rd closest individual	$\frac{2}{2!}\,m^3r^5e^{-mr^2}$	$[(3)(5)/2!]\,(1/2)^3\,\sqrt{\pi/m}$	$\frac{3}{m}$	$0.23883\frac{1}{m}$	$0.2717\frac{1}{\pi}$

and the variance is

$$\sigma_m^2 = \frac{m^2}{nN - 2} \tag{7.19}$$

Angle-order method

A combination of the angle and order methods (the angle-order method) is recommended for best results. Comparison of observed values of the mean and variance with theoretical ones in Table 7.2 should be made for non-random populations or when such is suspected. The angle-order method is based on the assumption that the area may be divided into several small fractions or sectors A_i in which the individuals will be distributed randomly or uniformly, even though individuals are aggregated over a large area A (Morisita 1957). Density may then be estimated from individual small fractions of the area. This approach is applicable to a mosaic of patches of different plant densities, but probably is not as useful when clumps are each composed of relatively few individuals. Sample points are randomly or regularly placed over an area, and the area around each point is divided into equiangular sectors. The distance r to the nearest nth individual, say $n = 3$, in each of four sectors k at N points is measured (Figure 7.4). Then, two methods of calculating the

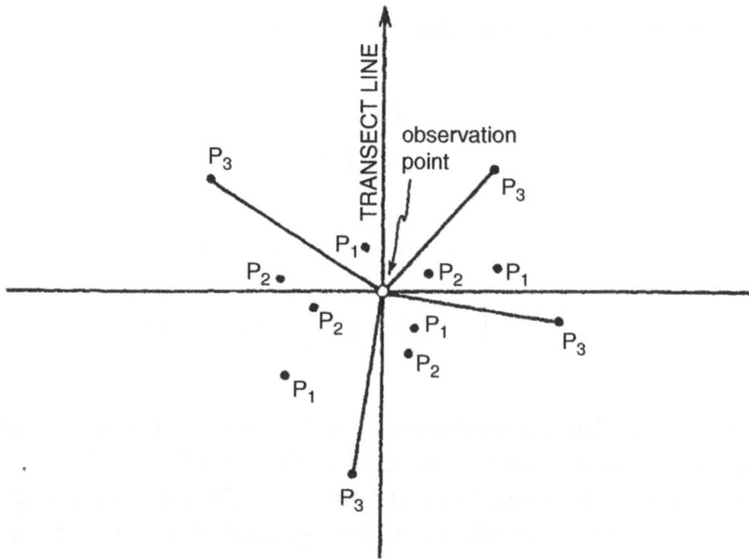

Figure 7.4 Angle-order technique where measurement of distance d is to third nearest plant in each quadrant from a point (P_i = plant order in each quadrant).

estimates for density are available. One involves the calculation of $1/r^2$ ($r = d$) for each sector and results in the estimates

$$m_1 = \frac{n-1}{N} \sum_{j=1}^{kN} \frac{1}{r_j^2} \tag{7.20}$$

$$\sigma_{m_1}^2 = \frac{1}{kN} \left[\frac{1}{A} \left(1 + \frac{1}{n-2} \right) \sum_{i=1}^{a} A_i m_i^2 - m^2 \right] \tag{7.21}$$

where m_i is the density estimate from distance measures from a small area or sector and m is the estimate of density over all points. The number of small areas (sectors) is indicated by a. The density estimate m does not depend on A_i or m_i, so these need not be known, as long as individuals are distributed randomly over A_i. The density estimate m in Equation 7.20 does not involve the area A_i that is only a fraction of A (total area covered by the vegetation type). Nor is the estimate affected by m_i, the mean density of the fraction area A_i. The other method to estimate density uses

$$\sum_{j=1}^{k} r_j^2$$

for each sample point, and then the density is estimated by

$$m_2 = \frac{nk-1}{N} \sum_{i=1}^{N} \frac{k}{\sum_{j=1}^{k} r_{ij}^2} \tag{7.22}$$

and its variance is

$$\sigma_{m_2}^2 = \frac{1}{N} \left[\frac{1}{A} \left(1 + \frac{1}{nk-2} \right) \sum_{i=1}^{a} A_i m_i^2 - m^2 \right] \tag{7.23}$$

The first method calculates the density in each sector (small area A_i), then averages these density estimates across all sectors. The second method calculates the average density at each sample point and then averages this estimate across all sample points. The density estimation method with the greater precision is determined by a comparison of $\sigma^2_{m_1}$, and $\sigma^2_{m_2}$. The first method is the best if there is a significant difference in density among sectors at a sample point. The second method is more accurate if the distribution over the entire area is random, and $n \geq 3$ and $k \geq 2$.

Some plant populations, however, have relatively uniform dispersal within a clump, rather than a random distribution. The sum

$$\sum_{j=1}^{k} r_j^2$$

is smaller for a uniform distribution than for a random one, and

$$m_1 < m < m_2$$

The best estimate of m is then obtained by

$$m = \frac{m_1 + m_2}{2} \tag{7.24}$$

if a hexagonal distribution is assumed. This estimate is used for random or uniform distributions, but m_1 should be used exclusively if $m_1 > m_2$. Note that in Equation 7.21, arithmetic considerations require that n be at least 3. It is recommended that $k = 4$ and $n = 3$. That is, use the distance to the third individual in each of four sectors. Morisita (1957) tested the angle-order method on artificial populations and found that density estimates for regularly, randomly, and non-randomly distributed populations were within 0.1, 1.8, and 5.0%, respectively, of actual density (as counted in quadrats).

In conclusion, the concerns with weaknesses in data collection from distance methods are not as important as the method used for data analysis and estimation of means and their variances. Plant ecology studies that do not adhere to sampling design methods and subsequent field applications may not result in adequate estimates of means and their variances for density.

7.5 Line transect

The probability of a circle (a plant) of a particular diameter being bisected by a transect at random is proportional to an area surrounding the transect with boundary at radius r distance from the transect line and its terminal points. The expected number of contacts for all plants of the same diameter is proportional to this area and the number of individuals in the class or species.

Chords of intersection are measured along a transect, and partial chords are measured only at one end of the transect. These chord lengths C are summed. An unbiased estimate of the number of circles (i.e., density) per unit area is

$$m = \frac{\sum C(1/D)}{nL\frac{1}{4}\pi \sum D} \tag{7.25}$$

where n is the number of random transects, each of length L, and D is the diameter of the plants (circles). Since the expected value of C is (1/4) D

$$m = \frac{(2/\pi)\sum(1/C)}{nL} \tag{7.26}$$

One modification of the above is the length of the longest chord parallel to the transect of an individual plant whose center is within a given radius of the transect. The equation for estimation of density from the longest chord parallel to the transect is

$$m = \frac{\sum(1/D)}{nL}\frac{1}{C} \tag{7.27}$$

Density can also be estimated from transect chords with supplementary data on longest chords by the following equation

$$m = \frac{k^2}{\bar{C}nL\sum D}\frac{\text{AM}}{\text{HM}} \tag{7.28}$$

where AM and HM are the arithmetic and harmonic means, respectively, of longest chords parallel to the transect of plants intersected by the transect and k is number of plants intersected. Density estimates for ellipses instead of circles used for plant shapes are obtained by

$$m = \frac{(2/\pi)\sum(1/C)}{\bar{C}nL} \tag{7.29}$$

Strong (1966) used the line-transect method to obtain density estimates that were improved over those suggested by McIntyre (1953). In the former study, measurements are made perpendicular to the transect across the widest portion of each intercepted plant and labeled D for diameter. The density estimate m then is

$$m = \left(\frac{1}{D}\right)\left(\frac{A}{T}\right) \tag{7.30}$$

where A is the unit area and T is the transect length. This equation includes a correction for size of plants since a plant with $D = 2$ would have twice the probability of being intercepted as a plant with $D = 1$. Then, D is the average diameter encountered along the transect. This method makes no assumptions regarding the shape of plants and does not require a subjective correction factor. The theory used to derive Equation 7.30 was based on the number of interceptions possible and their probability of being intercepted in a given area. Perhaps a more direct way, provided from the work of Lucas and Seber (1977), to estimate density is to measure the diameter of the intercepted plant canopy of a tree, shrub, or herb. If any part of the canopy is touched by the line, the maximum diameter of that plant is measured. Then

$$m = \frac{\sum_{i=1}^{k} (1/D_i)}{L} \tag{7.31}$$

where D_i, is the individual diameter of the ith plant, there are k plants intercepted, and L is the line length. The mean and variance are estimated from the usual equations given in Chapter 3.

7.6 Comments

Shrubs and small trees are often recorded in 16, 10, or 4 m^2 quadrats, and very few studies have been conducted in woody communities to determine the precision and efficiency of various quadrats. Herbaceous communities are most frequently sampled by 1 or 0.1 m^2 quadrats and plotless (distance) methods have become more widely used in forested communities. Generally, the size of the quadrat ultimately used is dependent on the community structure, plant morphology, and the purpose of the study.

Distance measurement techniques are among the simplest to use, provided the population consists of discrete individual plants. Quadrat boundaries are virtual and quadrat size is based on the population density under consideration. Generally, distance measures are most useful when applied to one species in autecological studies, but these measures can be used to study an entire community. The number of samples needed, however, to adequately sample the least abundant members of a community may preclude this type of sampling. Sometimes such community studies do not require the counting of an absolute number of individuals by species. Instead, an estimate of relative density may be used to contrast species composition between plant communities. Relative values, however, are dependent on the estimated density of all species within a community and may, therefore, be biased and misleading. When measurement of density is carried out within very dense or very sparse communities, the use of distance measures may be inefficient because considerable time may be spent on searching for the required individual plant. This problem may be more serious when measurements are made to the third or more distant plant, such as in Morisita's (1957) angle-order method.

If plotless methods are employed, distribution patterns must be assessed before the sampling method is chosen. For randomly dispersed individuals within a population, the PCQ method is the most reliable. Most populations are not randomly dispersed, but rather are contagiously dispersed. For non-randomly dispersed individuals, the angle-order method appears to be the most reliable.

Plotless methods require less time and may be more efficient for random populations. Randomness of plants may not always be the case, however, and appropriate techniques should be used in non-random distribution patterns. In fact, quadrat methods may integrate variations in pattern and departures from randomness better than plotless methods and may, therefore, be more precise in an estimation of density. In some vegetation types, quadrat methods may be inefficient and give inaccurate estimates of density. A detailed study is needed to determine the efficiency and precision of all sampling methods in a particular vegetation type before more specific recommendations can be made regarding which set of methods is appropriate.

The sample size needed is defined as the number of observations to estimate a mean (average) with a desired confidence level of probability. An adequate number of samples must be taken in order to obtain a reliable estimate of the true mean. An equation to determine an adequate size is given in Chapter 3 (Equation 3.43). Generally, the number of quadrats counted for density is a function of the variation among individual samples. In other words, the greater the variation, the greater the number of quadrats needed. There appears to be two major ways that sample size can be determined for density: the first is a function of the area to be sampled, and the second relates to the number of individuals counted per quadrat.

The ecologist might set a standard of 5 or 10% sampling intensity for an area. For example, an area of a particular vegetation type covers 10 ha and employs 5000 quadrats of 1 m^2. Alternatively, this area could be sampled by fifty 10×10 m plots.

Optimum sample size to minimize variance is not a function of the area sampled, but rather is a function of the number of samples containing individuals of a species. This is related to the spacing of individuals. Species that are in close proximity may be sampled with fewer quadrats or distance measures than species which are widely spaced.

7.7 Bibliography

Bartlett M. S. 1948. Determination of plant densities. *Nature* **162**: 621.

Bonham C. D. and Ward R. T. 1970. Phytosociological relationships in alpine tufted hairgrass (*Deschampsia caespitosa* (L.) Beauv.) meadows. *Arct. Alp. Res.* **2**: 267–275.

Bonham C. D., Reich R. M., and Leader K. K. 1995. Spatial Cross correlation of *Bouteloua gracilis* with site factors. *Grassland Sci.* **41**: 196–201.

Bormann F. H. 1953. The statistical efficiency of sample plot size and shape in forest ecology. *Ecology* **34**: 474–487.

Bourdeau P. F. 1953. A test of random versus systematic ecological sampling. *Ecology* **34**: 499–512.

Braun-Blanquet J. 1932. *Plant Sociology*. (Translated by G. D. Fuller and H. S. Conard.) McGraw-Hill: New York.

Cain S. A. 1936. The composition and structure of an oak woods, Cold Springs Harbor, Long Island, with special attention to sampling methods. *Am. Midl. Nat.* **17**: 725–740.

Catana A. J. 1963. The wandering quarter method of estimating population density. *Ecology* **44**: 349–360.

Dice L. R. 1948. Relationship between index and population density. *Ecology* **29**: 389–391.

Eddleman L. E., Remmenga E. E., and Ward R. T. 1964. An evaluation of plot methods for alpine vegetation. *Bull. Torrey Bot. Club* **91**: 439–450.

Evans F. C. 1952. The influences of size of quadrat on the distributional patterns of plant populations. *Contrib. Lab. Vertebr. Biol.* **54**: 1–15.

Greig-Smith P. 1983. *Quantitative Plant Ecology*, 3rd edn. University of California Press: Berkley, CA.

Hanson H. C. 1934. A comparison of methods of botanical analysis of the native prairie in western North Dakota. USDA 1. *Agric. Res.* **49**: 815–842.

Heady H. F. 1958. Vegetational changes in the California annual type. *Ecology* **39**: 402–416.

Lang G. E., Knight D. H., and Anderson D. A. 1971. Sampling the density of tree species with quadrats in a species rich tropical forest. *Forestry Sci.* **17**: 395–400.

Lucas H. A. and Seber G. A. F. 1977. Estimating coverage and particle density using the line intercept method. *Biometrika* **64**: 618–622.

McIntyre G. A. 1953. Estimation of plant density using line transects. *J. Ecol.* **41**: 319–330.

Morisita M. 1954. Estimation of population density by spacing method. *Kushu Univ. Fac. Sci. Me. Ser. E* **1**: 187–197.

Morisita M. 1957. A new method for the estimation of density by the spacing method applicable to non-randomly distributed populations (translation by USDA, Forest Service. 1960). *Physiol. Ecol.* **7**: 134–144.

Mueller-Dombois D. and Ellenberg H. 1974. *Aims and Methods of Vegetation Ecology*. John Wiley & Sons, Inc.: New York.

Persson O. 1971. The robustness of estimating density by distance measurements. In Patil G. P., Pielou E. C., and Waters W. E. (eds.), *Statistical Ecology, Vol. 2, Sampling and Modeling Biological Populations and Population Dynamics*. Pennsylvania State University Press: University Park, PA, 175–187.

Pollard J. H. 1971. On distance estimators of density in randomly distributed forests. *Biometrics* **27**: 991–1002.

Pound P. and Clements F. E. 1898. A method of determining the abundance of secondary species. *Minn. Bot. Stud.* **2**: 19–24.

Raunkiaer C. 1909–10. Formationsundersogelse og formationsstatistik. *Botanisk Tidsskrift* **30**: 20–132 (English translation Raunkiaer 1934 below).

Raunkiaer C. 1912. Measuring apparatus for statistical investigations of plant formations. *Svensk. Bot. Tidsskr.* **33**: 45–48.

Raunkiaer C. 1934. The life forms of plants and statistical plant geography (English translation Stearns R. W. 1959. Floristic composition as measured by plant number, frequency of occurrence, and plant cover. In *Techniques and Methods of Measuring Understory Vegetation.* (Proceedings of a symposium.) Southern Forest Experiment Station and Southeast Forest Experiment Station, 84–95.

Reich R. M. and Bonham C. D. 2001. Spatial analysis of grazed white bursage in the Lake Mead National Recreational Area, Nevada, USA. *Grassland Sci.* **47**: 128–133.

Stearns R. W. 1959. Floristic composition as measured by plant number, frequency of occurrence, and plant cover. In *Techniques and Methods of Measuring Understory Vegetation.* (Proceedings of a symposium.) Southern Forest Experiment Station and Southeast Forest Experiment Station, 84–95.

Strenge E. N. 1998. Spatial pattern analysis of vegetation and soil factors within the shortgrass steppe of North-central Colorado. M.S. Thesis. Colorado State University, Fort Collins, CO.

Strickler G. S. and Stearns F. W. 1962. The determination of plant density. In *Range Research Methods. A Symposium.* (Denver, CO.) USDA Forest Service Miscellaneous Publication No. 940, 30–40.

Strong C. W. 1966. An improved method of obtaining density from line-transect data. *Ecology* **47**: 311–313.

Walker B. H. 1970. An evaluation of eight methods of botanical analysis on grasslands in Rodesia. *J. Appl. Ecol.* **7**: 403–416.

8

Biomass

All biological activities of plants and animals are dependent on the energy of gross primary productivity. "Primary production" is defined as the energy fixed by plants, and it is the most fundamental characteristic of an ecosystem. Measurements of primary production are necessary for the proper understanding of ecosystem dynamics. Vegetation composition, based on biomass, is one of the best indicators of species dominance within a plant community. This chapter deals with measurements of primary production in terrestrial ecosystems. This vegetation measure is referred to as "biomass" in the discussion that follows.

8.1 Herbaceous biomass

The best and probably the most commonly used method for measurement of herbaceous biomass is to clip or harvest total standing biomass. Sampling techniques and methodologies have been developed to obtain estimates of herbaceous biomass and these methods are grouped into three categories: direct methods, indirect methods, and a combination of direct and indirect methods.

8.1.1 Harvesting

There is a variety of equipment to harvest herbaceous production; for example, grass and hedge shears, sickles, power mowers, and mowing machines. The choice depends largely on plot size, topography, and growth characteristics of the plants. However, native herbaceous vegetation is commonly clipped with grass shears and browse production by pruning shears. Grass shears are useful for clipping small quadrats and lawnmowers are useful only for herbage under 15 cm tall. Clipping height above ground level depends on the objectives of the study.

Weighing and drying harvested material

The weight of live plant materials includes inter- and intracellular water, as well as external moisture from vapor condensation, precipitation, and so forth.

Measurements for Terrestrial Vegetation, Second Edition. Charles D. Bonham.

Therefore, the weight of freshly harvested plant materials is highly variable and depends on the moisture status of plant, soil and atmospheric conditions. Then, for meaningful interpretation of results, biomass is expressed in terms of oven- or air-dry weight. In large-scale surveys, it is not feasible to keep all the harvested materials to determine the air- or oven-dry weights. Harvested materials are weighed fresh, or "green," in the field and recorded. A few samples of each species or species groups are air-dried. The ratio of dry to fresh weight is used to convert biomass data of respective species to dry weight. All herbage from sample quadrats may either be weighed together or separated into major species or species groups. Investigators often separate two or three dominant grasses, and lump the remaining grasses into a group, referred to as "other grasses." Forbs, shrubs, and trees can be treated similarly. Samples are to be dried within 24 h to keep fermentation and respiration losses to a minimum.

Herbage samples are oven-dried at 60 °C in scientific studies, but temperatures between 92 and 105 °C are also used. Lower temperature will prevent a reduction or loss of chemical constituents. A slight loss of ammonia may occur when young plant material is placed in an oven at 95–100 °C. Losses also occur in carbohydrates. A sample of each species should be dried separately and weights combined to obtain total biomass estimates from the vegetation type.

8.1.2 Measurement of biomass

Direct method

Harvesting or clipping is the most common direct method to estimate herbaceous biomass production. A plot of known dimensions is placed on a land area. Vegetation biomass is then harvested from the three-dimensional volume of the quadrat (height × width × length). Plant biomass that is not rooted in the quadrat but occupies space in the volume over a quadrat is harvested, while portions of plants rooted in the quadrat that do not occupy the quadrat volume are not harvested (Figure 2.13). Vegetation samples are clipped and weighed by species, aggregates of species, or life-form categories, and are standardized to oven- or air-dry weight. Materials should be clipped to ground level. Then, the amount of material available as forage for herbivore class can be partitioned by plant species.

The two main objectives of harvesting herbaceous biomass are to determine the amount of forage available for grazing by herbivores of all sizes, or to evaluate habitat conditions for vegetation. The advantage of harvesting is that data can be utilized for a variety of purposes. In particular, harvesting provides:

1. Availability of long-term records of vegetation production.
2. Variability in the contribution of individual species to total biomass production over time.

3. The relation of total plant yield of biomass to other factors such as soil nutrients and moisture, and plant cover, frequency, and density.

Sampling procedures often contain subjective decisions about inclusion of litter, certain vegetation parts, such as floral structures, and standing dead material. The observer must be consistent about decisions made and clip to ground level. Improper species identification will always be one of the most common errors made in determination of biomass production on an individual species basis. Means and variances for estimates of biomass obtained by harvesting are calculated by the usual methods presented in Chapter 3.

Weight estimates and double sampling

Biomass is clipped at a relatively small number of observation points taken at random from a large sample of visually estimated points. Two sets of data then become available: (1) a large sample that contains only observations by the visual method (guesses), and (2) within the large sample, a small sample that contains clipped weights of biomass in addition to the estimated weights. The small sample data set is used to calculate a regression to show the relation of clipped plant weight y to the estimated weight observed by visual estimation x. Estimates of x for the large sample are then used to obtain predicted biomass weights y. The unit clipped may be plots, branches, whole plants, or parts.

That is, an estimate of the clipped mean $\bar{Y}$ over all units, had all plots or units been clipped instead of visually estimated, is

$$\bar{Y} = \bar{y} + b(\bar{x}' - \bar{x}) \tag{8.1}$$

where $\bar{y}$ is the sample clipped mean and $\bar{x}'$ is the indirect (guesses) and $\bar{x}$ is the direct (clipped) samples, respectively, and b is the coefficient for least squares regression of $\bar{Y}$ on the difference $(\bar{x}' - \bar{x})$, i.e. b is the rate of change in $\bar{Y}$ given a change in this difference. The variance of $\bar{Y}$ (the corrected mean estimate of biomass from double sampling) is

$$v(\bar{Y}) = \frac{s^2_{y.x}}{n} + \frac{s^2_y - s^2_{y.x}}{n'} \tag{8.2}$$

where $s^2_{y.x}$ is the residual sum of squares divided by the number of double sample degrees of freedom $(n - 1)$

$$s^2_{y.x} = \frac{\Sigma^n[y_i - (a + bx_i)]^2}{n - 1}$$

and

$$s^2_y = \frac{\Sigma^n_{i=1}(y_i - \bar{y})^2}{n - 1}$$

where y_i is an individual unit of weight clipped and $\bar{y}$ is the mean for clipped weight. Say, for example, the sampling units are plots. x_i is the individual visual estimate of the individual clipped weight y_i. Then, the variance $v(\bar{Y})$ in Equation (8.2) is minimized by selection of n (the number of plots to be double sampled) and n' (the number of plots to be visually estimated only). This selection should always include costs to obtain both n and n'. Then, total cost C for double sampling is minimized by solving for a minimum C value when

$$C = nc_n + n'c_{n'} \tag{8.3}$$

where c_n and $c_{n'}$ are costs associated with a plot that is both visually estimated and clipped (n) and a plot that is visually estimated only (n'). Cochran (1977) shows that Equation (8.3) and

$$\frac{n}{\sqrt{V_n C_{n'}}} = \frac{n'}{\sqrt{V_{n'} C_n}} \tag{8.4}$$

can be used to determine n, and n' as an optimum combination. As an example, let the variance for clipped plots V_n be 36, and cost C_n equal unity, and the variance for a regression double-sample estimate $V_{n'}$ be 169 with a cost $C_{n'}$ of 20% of clipped plots. That is, it costs five times more to clip than to visually estimate a plot. Equation (8.4) is solved by

$$\frac{n}{\sqrt{36(0.2)}} = \frac{n'}{\sqrt{169(1)}} \quad \text{or} \quad \frac{n}{n'}\frac{\sqrt{7.2}}{\sqrt{169}} = \frac{2.7}{13} = 0.21$$

Then, Equation (8.3) is used to find n and n'. That is, if 100% is available for total costs C and $C_{n'}$ is one-fifth the cost of $C_{n'}$, substitution into Equation (8.3) gives

$$100 = n + 0.2n'$$

Dividing through by n' gives the specific costs and variances expected

$$\frac{100}{n'} = \frac{n}{n'} + \frac{0.2n'}{n'} = 0.21 + 0.2$$

or

$$n' = \frac{100}{0.41} = 244 \quad \text{and} \quad n = (244)(0.21) = 51$$

A ratio of 4.8 (244/51) or 5:1 is then needed to estimate total biomass within the specific costs and variances expected. The ratio is also determined from $n/n' = 0.21$

and $1/0.21 = 4.8$ as before. If Equation (8.1) is a close fit (i.e., if visual estimates are close to clipped values), then this ratio will decrease. For instance, if the variance from the regression estimate $V_{n'}$ is 72 instead of 169 and the cost ratio remains the same, then

$$\frac{n}{n'} = \frac{\sqrt{7.2}}{\sqrt{72}} = \frac{2.7}{8.5} = 0.32$$

and the ratio is 3:1. So an increase in ability to guess close to clipped weights of biomass reduces the need to clip, since both provide essentially the same information. Another interpretation is that as one misses the clipped values by visual estimates, the variance from regression $V_{n'}$ increases and the ratio of n' to n increases because more visual estimates are needed to adequately estimate the variance of biomass by visual estimates. However, a combination of variance estimates from clipped and visual estimates will usually result in a more efficient sampling estimate. This is especially true when $1/n$ is small (i.e., the reciprocal of the number of clipped plots is small), which implies that a large number of plots would have to be clipped to permit adequate estimation of the mean and its variance for biomass, unless double sampling is used. Since an adequate number n for double samples is needed to estimate the regression coefficients, a minimum of 25–30 such observations should be made for large areas (several hundred hectares) of homogeneous vegetation composition, that is, a vegetation type. Then, the ratio is used to determine the number n' of plots to be visually estimated. For the first example, a ratio of 5:1 was needed, and if n must be at least 25, then $n' = 125$. That is, 25 are both clipped and visually estimated for biomass.

Double sampling then enables one to make more precise estimates of the average biomass weight. Since both large (visually estimated) and small (clipped and visually estimated) samples are taken by randomization, an error variance can be calculated for the regression estimator and also for estimated plant biomass.

8.1.3 Non-destructive methods

Measurements of vegetation biomass at a given point in time by harvesting techniques have some limitations. For example, some ecological studies warrant repeated measurements, within the same growing season, of permanently located sampling units, and this is not possible with a direct harvest method. The only alternative is to select a suitable non-destructive method for estimating biomass production. It was previously noted that harvesting is time consuming and expensive. Therefore, indirect methods are useful since they eliminate or reduce clipping and can be used to obtain a measure of biomass. These methods use ratio or regression estimators, wherein certain easily measured vegetation characteristics are mathematically correlated to biomass production. Vegetation characteristics, such as leaf length, stem length, number of stems, crown diameter, basal diameter, and cover, all have a degree of correlation with biomass.

Regression equations are generally of the form

$$y = a + b\,[f(x)] \tag{8.5}$$

where y represents the biomass estimate for each species, a is a constant, b is the slope of the line, and $f(x)$ is a function of the measured plant characteristic. The $f(x)$ may be log x or other relationships that need to be determined for any characteristic such as cover for each species. Equations are developed for individual species. However, data can be combined from different vegetation areas in the same locale to develop the equation for each species. Data collection procedures must be kept as consistent as possible with respect to both biomass production and its correlated characteristic. Otherwise, variability associated with data collection is increased and the equation is less precise for the estimate of biomass.

Spectral reflectance

Spectral responses of objects indicate the patterns of reflected or emitted energy that is derived from absorbed solar energy. Because plant species have different spectral responses according to biomass concentrations, it is possible to obtain estimates of biomass from these responses. Specific bands of the spectrum must be matched to various levels of standing vegetation biomass. Once this is accomplished, spectral analysis can be used to measure and monitor the plant biomass of an area. Radiometers are used in field measurements of biomass when non-destructive estimation of biomass is needed.

Pearson and Miller (1972a) developed a method using a radiometer to estimate the biomass of a shortgrass prairie site. They used a 0.25 m^2 plot, obtained spectral reflectance curves, clipped small portions of the plot, and again obtained spectral readings. The process was continued until the plot was clipped to ground level. Naturally, curves varied according to the amounts of green vegetation present in the plot, and curves were predictable. In particular, it was noted that an inverse linear relationship existed between total grass biomass and spectral reflectance at a band of 0.68 μm ($R^2 = 0.70$). However, when all plant biomass was included, a direct linear relationship was found at a band of 0.78 μm ($R^2 = 0.84$). All measurements were obtained with a portable radiometer placed approximately 1.0 m above the plot.

Ratios of spectral bands are also used to predict green standing biomass of vegetation. Ratios from 0.78/0.68 μm to 0.800/0.675 μm have been used successfully ($R^2 > 0.90$) to estimate green biomass (Pearson and Miller 1972a, Waller *et al.* 1981, Boutton and Tieszen 1983). These workers have also noted that time of day affects the precision of biomass estimates because of changes in amount of sunlight falling on the plots. Tucker (1980), in particular, noted that readings for spectral reflectance must be made in direct sunlight and within one to three hours of noontime. Otherwise, variability caused by the solar zenith angle limits precision of biomass estimates obtained from radiometers. Significant errors also accrue when green vegetation is less than 30% of the total vegetation biomass in the area. It follows that the method is more reliable during the active growth of vegetation. On the other hand, the use of radiometers over large areas is advantageous because measurements can be obtained in a matter of seconds from either high altitude aircraft or spacecraft.

Growth form and production

Some vegetation ecologists have hypothesized that within a given species, volume–height or height–weight distribution is more or less constant. In other words, a species growth form is relatively constant. However, height–weight distribution of biomass of a species may vary because of major differences in amount of leaf material, for example, in a drought year. If no variations in growth form occurred, then it would be easy to determine volume or weight of plants from their height. A number of attempts have been made to correlate height of plants with their weight or volume. In fact, growth forms of species vary so much according to weather conditions, stand density, defoliation, and available mineral nutrients that it is difficult to use height alone to predict weight or volume of plants. Plants grow taller in more favorable years or in areas with high densities and have a more cylindrical or conical shape than under other conditions.

The relationship between height and herbage volume (or air-dry weight) shows that in most species more volume is concentrated near ground level. As heights of plants increase, biomass volume in grasses usually decreases (Clark 1945, Heady 1950, and Husch *et al.* 2003). Growth form for the same species varies markedly between plants from different years, from different elevational zones, and from different sites (Table 8.1) (Clark 1945).

Kelly (1958) measured leaf and shoot height to predict production. The effect of time within a growing season affects the relationship between height and biomass of a species. Then, a separate term in the equation is needed to account for seasonal differences. For example, the equation for predicting weight of biomass from plant height is

$$y = a + bx \tag{8.6}$$

where y represents biomass amount, a is a constant, and b is the rate of change in biomass amount per unit of change in height x. If seasonal effects exist, then

Table 8.1 Percentage of biomass removed at different percentage heights above-ground level for grasses collected in the same year from the Oakbrush (*Quercus*), *Aspen* (Populus) – Fir (*Abies*), and Spruce (*Picea*) – Fir (*Abies*) Zones. (Republished with permission of Society of American Foresters, from Variability in growth characteristics of forage plants on summer range in central Utah. Clark, I. 1945. *J. Forestry* **43**(4): 273–283; permission conveyed through Copyright Clearance Center, Inc.)

Height	(*Bromus marginatus*) Mountain brome			(*Agropyron trachycaulum*) Slender wheatgrass			(*Poa pratensis*) Kentucky bluegrass		
percent	Oak	Aspen	Spruce	Oak	Aspen	Spruce	Oak	Aspen	Spruce
100	0	0	0	0	0	0	0	0	0
95	–	–	2	1	1	1	1	1	1
90	1	1	6	3	3	3	3	2	4
85	2	–	9	5	5	5	4	3	9
80	3	2	13	7	7	7	7	4	14
75	5	3	16	10	9	9	9	6	18
70	7	4	20	12	12	11	11	8	22
65	10	5	24	15	15	14	13	10	26
60	13	7	28	18	17	18	15	12	29
55	17	9	32	21	21	21	18	15	33
50	21	12	37	25	25	25	21	18	36
45	26	15	41	29	29	30	24	21	39
40	32	19	45	34	35	33	28	25	42
35	39	24	50	39	39	39	33	28	46
30	46	30	56	45	45	45	38	33	51
25	55	37	64	51	52	52	44	38	56
20	63	46	69	59	61	60	52	46	61
15	72	63	76	69	68	69	62	56	70
10	82	71	83	78	82	79	74	69	82
5	92	86	93	89	90	90	87	84	91
0	100	100	100	00	100	100	100	100	100

Equation (8.6) includes another term for this effect z and its associated rate of change coefficient (c) as in

$$y = a + bx + cz \tag{8.7}$$

These equations apply to any situation where variables, other than height or season, are used to predict biomass amount of a species or group of species. Equation constants and coefficients are estimated from least squares procedures as used in regression analyses.

Cover method

Some studies show that cover is a suitable measure for prediction of biomass production of some species (Anderson and Kothmann 1982). If this is so, use of cover values as an estimate of herbage production can save a great deal of field and laboratory time. Cover–weight relationships are developed by measuring cover and fresh or oven-dry weights of species in a number of randomly located quadrats. The relationship between cover and oven-dry weight of a species is much more easily obtained than that of cover and fresh weight. A simple linear relationship often exists for the former relationship. The data are then used to compute cover–weight relationships for each species through regression analysis. A high positive correlation between ground cover and vegetation production often exists.

Photographs of vegetation may be used for measuring vegetation cover. Cover is then used to predict biomass. Wimbush *et al.* (1967) used vertical color stereophotography to record cover changes in alpine flora. Wells (1971) modified this technique and used two 35 mm cameras fired simultaneously to take stereophotographs at approximately chest height. The transparencies are viewed directly by transmitted light under a zoom stereoscope, and plant cover is measured by point counts made on photographs with the aid of a counter. The photographic approach to measure cover can expedite the procedure to develop cover–weight equations to estimate biomass. A photograph of a quadrat is taken before clipping and weighing the vegetation. Photographic cover and oven-dry weights are then used to solve regression equations. Thus, cover is a useful method for prediction of oven-dry weights of species, but not fresh field weight.

Density method

Studies have been conducted to estimate herbaceous production from plant density, but results have not been satisfactory. All plants growing within a plot are counted to obtain an estimate of plant density. Biomass is then calculated by multiplication of the average plant weight by the average density. The density–weight method may give estimates of herbaceous production that are several magnitudes of those estimated by clipped plots. This is due, in part, to the bias introduced by the nonrandom distribution of plants and considering only "rooted" plants within plots.

Point method

The point-frame method is extensively used for measurement of vegetation cover. If at each sampling location, species biomass is also determined simultaneously, then a regression equation can be used to estimate biomass by point sampling. After the pins in a point frame have been lowered and all hits recorded, a quadrat is lowered over the point frame and all vegetation is clipped, dried, and weighed. Hughes (1962) estimated herbage production by use of an inclined point frame that had ten pins and a 0.9 m^2 quadrat. Both plant cover and frequency data recorded by the point method gave reliable estimates of biomass production. The use of a small quadrat will give a small range in biomass relative to cover contacts by pins. In turn, higher precision is attained for biomass estimates.

Biomass prediction from multiple factors

A single factor such as plant height or cover may not be reliable as a predictor of herbage yield. In fact, as previously mentioned in Equation (8.7), a combination of variables usually provides a more precise estimate of biomass production than a single variable. Variables can be added, divided, or multiplied in order to find an acceptable equation. Hickey (1961) found that compressed crown diameter and compressed leaf length, all combined, gave best estimates of production. Compression is accomplished by squeezing herbaceous plants between the hands.

Biomass yield can be estimated (Teare and Mott 1965) by regression equations based on leaf-area index (LAI) and longest point of the plant (LP). The mathematical product of LAI and LP is often the only important term in the equation. However, another component consists of leaf and stem material, so both the product LAI × LP or LP alone may be used in the regression equation, but not both because of correlation between these two variables.

Andariese and Covington (1986) estimated biomass from logarithmic regression equations to relate above-ground biomass to basal area for each species. Logarithmic equations are discussed further in Section 8.3. The ability to predict plant biomass from basal area has an advantage over some other indirect methods, such as the weight-estimate method, because personal bias is reduced.

Meteorological data method

Precipitation and air temperature have a direct effect on plant growth. Various attempts have been made to develop models to predict herbage biomass production from precipitation and other data. The model utilizes soil moisture, precipitation, mean air temperature, and solar radiation as input data to calculate the ratio of actual transpiration T to potential transpiration T_p as a yield index. Two-thirds of the field-measured biomass yields were within one standard deviation of the forecast yields for April, May, and June. Total biomass yield is often closely related to total precipitation in a given month (Wight and Neff 1983, and Wight *et al.* 1984).

Wisiol (1984) chose six equations to estimate grassland yield for specific years for a variety of sites. The climatic and biomass production data used to test the models came from 28 locations in North America, Asia, Africa, Australia, New Zealand, and Europe. The smallest error was obtained with the Sahel-Sudan equation and each equation estimated yields within 5% or less at times. These findings show that models based on climatic data should only be used in the area for which they are originally developed.

8.1.4 Productivity

An analysis of ecosystem processes, such as seasonal patterns of biomass dynamics, nutrient content, consumption, and decomposition, is dependent on a precise estimate of net primary production. Above-ground plant material is usually considered as a measure of above-ground net primary production (ANP), and is estimated by harvesting standing crop when the system has apparently attained maximum standing crop. Productivity is the rate of biomass produced per unit time interval, while production is the amount of biomass accumulated over time. Standing crop, therefore, is weight of plant material present in a system at any given point in time. However, by the time measurements are taken, some material may have already senesced, detached, and disappeared from the observation point. The problem can be overcome by cumulative sampling. Cumulative sampling procedures utilize extensive seasonal sampling, and production is determined from a summation of maximum biomass estimates or positive changes in biomass. While this approach to estimation of production increases field work, it will increase precision of the estimates.

Singh *et al.* (1975) compared several methods of calculating ANP. They found that most methods used in computing net aerial production were correlated with each other; methods that yielded highest or lowest estimates differed from site to site. Some methods appeared to be better discriminators of perturbations on vegetation, whereas others proved more useful for distinguishing sites or years. No single method is universally applicable for estimating ANP.

Yiruhan *et al.* (2011) presented results of a relatively small, but important study on the amount of above-ground plant biomass material available to domestic goats. Two 50 × 50 m permanent plots (labeled S and G for easy reference) were established in a mountainous shrubland in China. Goats could enter freely from the outside and feed in the plots. The survey was conducted annually with a single replication from November 2004 to November 2008, a period of four years.

Plant mass was estimated using two methods: (1) harvesting all herbaceous plants and shrubs with a height of <1 m in a given area and (2) allometric equations were used to estimate plant mass from the breast height diameter (dbh) of shrubs with height >1 m. For the first method, a 50 m line was placed in both the S and G plots and ten 1 × 1 m frames were located every 5 m of each line. For four of the eight sampling dates (November and December 2004, October 2005, July and December 2006, April and July 2007, and November 2008) plants occurring within each frame were clipped at ground level. Plants were classified into five categories: grasses, forbs, ferns, short shrubs with a plant height <1 m, and standing dead material for four of the eight sampling dates. Fresh weights were determined for each category. A portion was oven-dried at 70 °C for > 48 h and used to determine oven-dried weight for each category. The total oven dried plant mass, including shrubs <1 m height without standing dead mass, was determined for each sample date.

Table 8.2 Allometric equations ($y = ax^b$) describing the relationship between the weight of a plant part (y, kg) and the diameter at breast height (x, cm)

Season	Plant part	*a*	*b*	R^2
Dry	Leaf	0.0295	1.635	0.768
	Xylem	0.0932	2.584	0.854
	Total	0.1332	2.363	0.863
Rainy	Leaf	0.0121	2.973	0.863
	Xylem	0.0144	3.122	0.842
	Total	0.0269	3.060	0.870

a and *b*, regression constants. Regression analyses for the dry (September–February) and rainy (March–August) seasons were performed based on the data in October 2004 and June 2005, respectively.

For the second method, allometric equations were developed for both the dry and rainy seasons to predict shrub mass for plant height >1 m. Solutions for these equations were based on data gathered outside the two, 5 × 5 m, smaller permanent quadrats. The dbh of each shrub with heights >1 m in the permanent quadrats was measured with a tree caliper along with leaf, xylem, and total mass. The dbh was measured at a height at which the stem diameter was constant.

Allometric equations were solved by using simple regression analysis (Yiruhan *et al.* 2011). Table 8.2 provides the results of applying the allometric equation, $y = ax^b$, to estimate plant masses where x and y are dbh (cm) and plant mass (kg), respectively. Regression coefficient values, a and b, are given for each plant mass category.

8.1.5 Considerations of herbaceous biomass sampling

Plot sizes and shapes

An investigator frequently encounters the problem of choosing the best size and shape of a plot because such dimensions may influence estimates of biomass production. The main criteria for optimum plot size and shape are to obtain minimum variance and an unskewed data distribution. Plots should also be convenient to use in the field. In general, from a theoretical point of view, a much greater precision is achieved by making the plots as small as possible and by increasing the number of sampling units. There are, however, disadvantages in reducing the size of the plot below a certain limit. In large-scale survey work it would be prohibitive to select, demarcate, and collect data from a large number of plots. For small plots, a greater portion of edge per unit area increases the difficulty of deciding which plants to include in the sample (Morris 1959). This bias is usually toward overestimation of the mean production value. The bias is more serious where plants are larger and fewer because incorrect inclusion or exclusion of a single individual influences total production to a greater extent than where plants are smaller and more numerous (Wiegert 1962).

Plots that are too small may also bias the observer's ability to identify the contribution of infrequent species unless a large number of observations are made. This bias occurs when dominant species occupy the entire plot. In contrast, plots that are too large result in excessive expenditure of time for a relatively small return of additional information. Increasing the size of plots reduces variance, and, thus, only a few plots would be included in the sample. Such an increase in plot size will result in inflation of the confidence limits of the mean because of rapid increases in the "student's t–value" for a few degrees of freedom. Thus, the advantage of increasing the plot size to reduce variance is offset if the sampling scheme involves only a few plots.

The problem then is to find an optimum size of plot to sample a given vegetation type without having an adverse effect on precision as a result of an increase in size of the plot and reduction in the number of sampling units. Even when the optimum size of the plot has been determined, a change in the shape of the plot may affect sampling error. Methods to find the optimum size and shape are discussed later in this section. Except for research work, it may not always be feasible to spend time and money to find an optimum size and shape of a plot. Selection and use of a plot size and shape usually reflects the experience of an ecologist in a similar vegetation type.

Theoretically, long, narrow plots have smaller coefficients of variation when placed with their axis parallel to the gradient of heterogeneity (Christidis 1931, Bormann 1953). However, if the direction of variability in soils or vegetation cannot be determined, then square or circular plots should be used because vegetation patterns expand in all directions from a sample point anyway.

Statistical errors remain constant or decrease as the amount of biomass increases and result in smaller variances for larger plots. The coefficient of variation for bunchgrasses is usually much lower than that for sod-forming grasses (Van Dyne *et al.* 1963). If the biomass material being sampled is fairly uniform, there will be little gain in precision by increasing the size of the plot.

Unless sampling areas are the same for plots varying in size, no valid comparisons can be made, as pointed out in previous chapters. Yet, there is value in listing ranges in plot sizes and shapes used in various vegetation communities. For example, Campbell and Cassady (1949) used a range of plot sizes from 0.4 to 9.3 m^2 and their statistical analysis would not have been meaningful because the sum of the area sampled by each plot size/shape was not equal among the combinations of plot sizes and shapes. There are many such studies reported in the literature that committed the same error and any recommendations as to plot size and shape to use should be disregarded.

In southern United States forested grasslands, a square plot of 1 m on a side is about the most convenient shape to use. On the other hand, circular plots are difficult to delineate in tall grass types. A square or rectangle with one side removed is easy to use in all types of vegetation, particularly in tall brush or tall grass types. In large areas of low precipitation and supporting low-growing plants, circular or long and narrow plots are thought to be generally satisfactory for biomass measurements. But again, increasing sample size overcomes the effects of spatial scales on mean and variance estimates of biomass. Economics, however, enters into selecting long narrow plots or a large number of small plots.

Determination of minimal sampling unit area

The minimal sample area can be determined from a system of nested plots (Figure 8.1). Initially, a small area, say, 10×10 cm ($0.01\ m^2$), is laid out and total

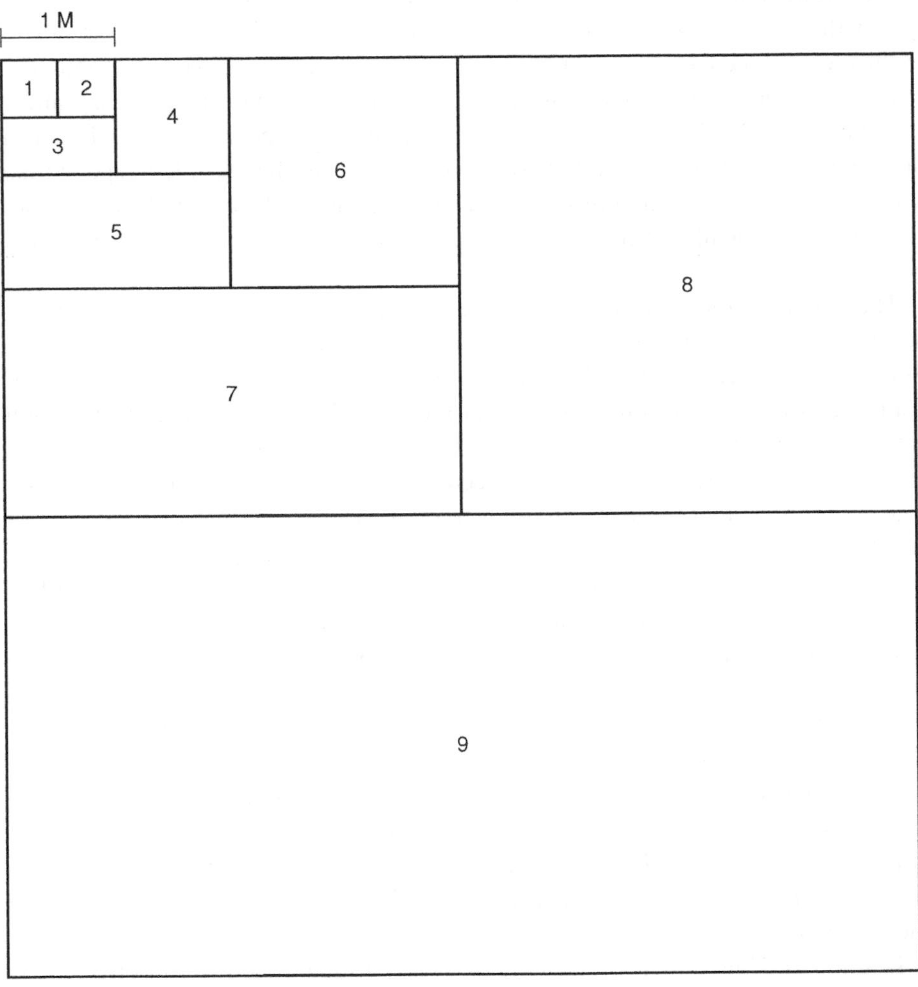

Figure 8.1 A system of nested plots for establishing minimal area. Each sub-plot numbered consecutively includes the area of the previous sub-plot. Thus, even-numbered sub-plots are square, and uneven-numbered ones are rectangular. (From Mueller-Dombois D. and Ellenberg H. 1974. *Aims and Methods of Vegetation Ecology*. © John Wiley & Sons, Inc. Redrawn with permission.)

biomass is clipped. Then the sample area is progressively enlarged to twice the size, to four times, to eight times, and so on. The additional biomass and time required to clip it is recorded separately for each enlarged area. The biomass, fixed cost (travel between locations, weighing, etc.), cost to clip each sub-plot in terms of time spent on clipping, and variance changes are recorded. The underlying principle used to estimate optimum plot size is minimization of total cost for a given variance, or alternatively minimization of variance for a given total cost. See Section 3.3, "Sample Size" for equations and Chapters 3 and 4 in Morrison *et al.* (1995), for nested quadrats.

If the total cost of sampling equals C, then the cost of a single plot equals C/n, where n is the number of sample observations (plots). The cost C/n consists of a fixed cost, Cf (which is independent of the size of the plot and again, as previously noted, involves travel between locations, weighing, etc.) plus X times a cost C_v that represents time spent on collection of data from a plot. The relative cost C_r equals cost for a given plot size relative to cost for the smallest plot size.

8.2 Shrub biomass

Shrubs constitute an important part of many ecosystems and provide browse, not only for wild animals, but also for various categories of domestic livestock. In fact, browse may constitute the bulk of available forage in some ecosystems. In other ecosystems, it may be the only available forage during certain times of the year.

A precise estimate of shrub biomass is important for an evaluation of the productivity of an ecosystem, cycling of nutrients and energy. An estimate of shrub biomass is also required to describe the amount of fuel for firewood, and appraisal of flammability. To evaluate shrub flammability, it is necessary to know not only shrub biomass weight, but also its size and distribution over an area. The weight of foliage is also useful for an estimation of transpiration use of water by shrubs.

Browse measurement is one of the more difficult components of vegetation to determine. Similar to the measurement of herbaceous production, many techniques have been used to measure quantities of browse. All are based on one of three standard procedures: direct harvest method, indirect weight estimates, or a combination of harvest and estimates. Clipping and weighing is generally the most precise method, but is also laborious. A large number of plots are often necessary to measure shrub production. Indirect methods for reliable estimates may involve less time than direct harvesting and permit a large number of observations to be obtained at a relatively low cost.

8.2.1 Non-destructive methods of shrub biomass

Limitations of direct harvest methods to obtain efficient estimates of shrub production led to development of indirect or non-destructive methods for production.

Measurements of crown area, circumference, diameter, and volume; plant height; basal stem diameter; twig diameter, length, and weight; and width of xylem rings, alone or in combination with other variables, have been used to estimate shrub production. The indirect measurement techniques can be divided into two broad categories: reference unit technique and dimension analysis. The reference unit technique involves matching a known unit of weight against observations. In the dimension analysis technique, a mathematical relationship is developed between easily obtained plant dimensions and foliage weight. Specific plant dimensions useful in estimating biomass of shrubs are discussed individually.

Reference unit technique

A small unit of plant such as a shoot of a given dimension is designated as the reference unit. The size of the reference unit should be 10–20% of the foliage weight of the average plant (Andrew *et al.* 1979, Kirmse and Norton 1985). A few reference units are clipped, and average green or dry green weight is determined. The next step is to count or estimate similar reference units. The number of estimated reference units is multiplied by the average weight of clipped reference units to estimate shrub biomass production. The actual sampling is preceded by a short period of training. During training the technicians familiarize themselves with a reference unit and the number of such reference units on different plants. Estimates are frequently checked with actual weights. The reference unit technique is not suitable for shrubs which have a compact, dense, unsegmented growth form. The technique was tested by Andrew *et al.* (1981) on two small shrubs in Australia and compared favorably with other techniques for estimating shrub biomass.

Kirmse and Norton (1985) compared the reference unit method to dimension analysis methods for two large shrubs in Australia. The amount of biomass variation accounted for by the reference unit approach ranged from 89 to 98%. The R^2 values obtained in applying the dimension analysis method were 0.937 and 0.948. The two methods took about the same amount of time. Therefore, Kirmse and Norton recommended the reference unit method for field application, rather than dimension analysis.

Crown area

Some studies have shown high correlation between the dimensions of crown area and leaf biomass. For example, correlation of shrub biomass production with ground cover was studied in Nevada. Crown area, length of the current year's twig elongation, and production were measured for several consecutive years in plots 5 m^2 in area (0.6 × 7.6 m). Production was determined by clipping varying percentages: 25, 50, and 75% of current annual growth was removed in spring; 10–30% in summer; and 40, 60, and 90% in winter. Highly significant correlations were found between spring twig growth and production. Thus, an estimate of production from the current year's twig length would be more precise from measurements made in spring. Crown cover was significantly correlated with production Y ($r = 0.4$, $Y = 40.0 + 9.8\,x$) and indicated that winter was the best time to estimate yield from shrub ground cover.

Medin (1960) used crown diameter to estimate shrub production on different sites in northwestern Colorado. Crown diameter of all plants was measured on each 40 m^2 plot. One plant per plot was randomly selected and clipped to obtain air-dry weights for all current annual twig growth after cessation of annual growth. The regression of production with crown diameter yielded a correlation coefficient of $r = 0.84$ (log weight = 1.25 + 1.66 log of crown diameter). Harniss and Murray (1976) used crown circumference C times height H to develop the mathematical model

$$DW = 0.0167HC^{1.25} \tag{8.8}$$

to predict the dry-leaf weight DW of shrubs. The model shows the predictor in three-dimensional form and model was tested at two different sites and it accounted for 93 and 80% variation, respectively. The results suggest that the geometric shape of the predictor is representative of height–circumference/weight relationship in big sagebrush and that the predictor may be used for other stands.

Murray and Jacobson (1982) used dimension analysis for predicting biomass of several shrubs. A simple linear model of the following form gave the best biomass predictions:

$$1n\,Y = 1n\,a + b\,1n\,H + c\,1n\,C \tag{8.9}$$

where Y is the weight (g), H is the height (cm), C is the circumference (cm), and a, b, and c are the Y intercept and slope coefficients, respectively. This model worked well in prediction of not only the leaf biomass but also live and dead twig biomass components. Note that Equation (8.9) is the log transform of Equation (8.8).

Crown volume

Crown volume has been shown in many studies to be an adequate predictor of the total leaf biomass of shrubs. At least two measurements (diameter and height) of the crown are required in addition to foliage weight to develop volume–weight relationships. In the case of shrubs with an irregular outline, an average of two diameter measurements is required to calculate volume. One diameter is recorded in the longest direction of the canopy and the other in the perpendicular direction to the first measurement. The calculation of volume depends on the geometric configuration of the shrubs. That is, shapes of conical, cylindrical, spherical, and so forth are considered. Some canopy shapes and geometric formulas to fit these shapes are given in Figure 8.2.

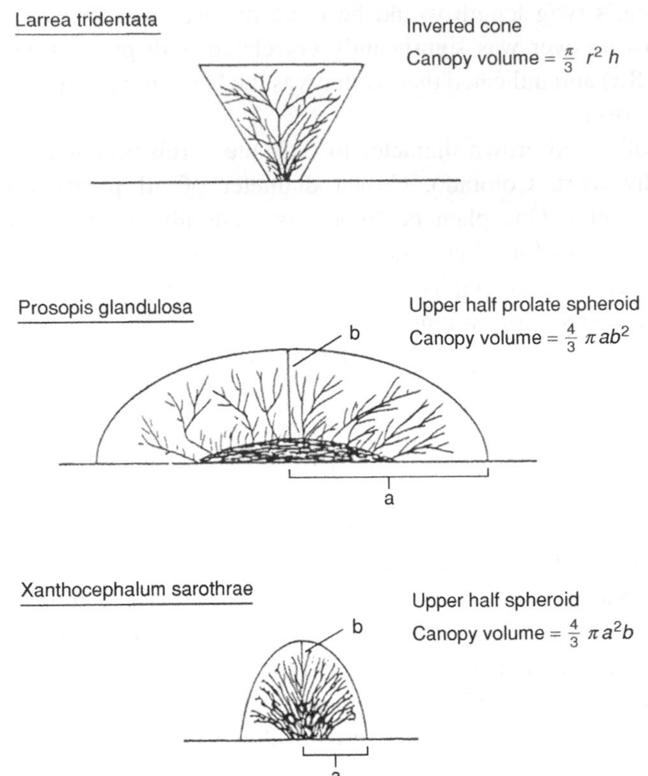

Figure 8.2 Canopy shapes and equations to describe shapes. (Reproduced with permission from Size-biomass relationship of several chihuahuan desert shrubs. Ludwig, J. A., J. F. Reynolds, and P. D. Whitson. 1975. *The American Midland Naturalist* **94**: 451–461. University of notre Dame.)

The crown volume–weight relationships as shown in mathematical models sometimes are developed for individual species and may be applicable only to areas in which these models are developed. An important consideration in computing crown volume is the openings in the canopy, as found by Rittenhouse and Sneva (1977). These authors found that canopy openings less than 30 cm were considered as a continuous intercept. Their mathematical models to predict the leaf and woody biomass of shrubs from height and two measurements of crown width had R^2 values that ranged from 0.72 to 0.95 for different models.

In western Montana, Lyon (1968) estimated shrub twig production from crown volumes on 11 different sites. Shrub crowns were measured in the summer while plants were in full leaf. One measurement was made through the long dimension of the canopy and another at right angles (eclipse diameters a and b, respectively). Height h was measured from ground surface and volume was calculated using

$$V = (\pi/4)\, abh \tag{8.10}$$

Current annual growth on each plant was clipped after leaf fall to obtain an estimate of twig production. Total annual growth was collected for plants with volume less than 2 m^3, while a random quarter of the shrub canopy was clipped for larger shrubs and the other three quarters were estimated. Results of regression analysis indicated that twig production on individual sites can be predicted precisely. Over 80% of the variation in twig production was associated with shrub volume for 6 out of the 11 sites. It was noted that prediction of biomass was least precise where soil surface disturbance had occurred recently. Therefore, unnatural factors for a site should always be noted, and plants should be measured by strata when possible. That is, regression equations should be developed for each uniform environmental condition present, to obtain the best results.

Uresk *et al.* (1977) used a double-sampling procedure to estimate biomass of different shrub components of sagebrush (*Artemisia tridentata*) in the state of Washington (USA). The following measurements were recorded for each plant: (1) longest diameter of the canopy, (2) longest diameter of the canopy measured at right angles to the previous dimension, and (3) maximum height. Shrubs were cut at ground level, and oven-dried weights were obtained for various biomass compartments (leaves, woody material, etc.) separated by hand. The same data were visually estimated on a large number n' of randomly selected sagebrush plants. Data from clipped shrubs n were used to adjust the visual estimates through linear regression. A separate equation was developed for each harvest category (leaves, wood, etc.).

One objective of the study was to minimize variance of estimated mean phytomass for each category for a fixed cost. It was assumed from experience that a clipped estimate of biomass was 120 times as expensive to obtain as were the dimension measurements. The optimum ratio of n' to n was estimated using the technique described by Cochran (1977), as in Equation (8.4) (above)

$$\frac{n}{\sqrt{V_{nCn'}}} = \frac{n'}{\sqrt{V_{n'Cn}}}$$

where V_n and $V_{n'}$ are the clipped and guessed biomass weight variances, c_n and $c_{n'}$, are the costs associated with the respective sample sizes n and n'. The objective is to find an optimum number of n to n' measurements.

Generally, a stepwise regression is used and volume is entered first to estimate the biomass of interest. The double sampling (ds) estimate is obtained by Equation (8.11), which has different notation

$$\operatorname{var}\left(\overline{Y}_{\text{ds}}\right) = S_{yx}^2\left[\frac{1}{n} + \frac{\left(\overline{X}_{n'} - \overline{X}_n\right)^2}{\sum_{i=1}^{20}\left(\overline{X}_{i'} - \overline{X}_n\right)^2}\right] + \frac{S_y^2 - S_{yx}^2}{n'} \qquad (8.11)$$

where $\overline{X}_n$ is the mean volume based on $N = 20$ clipped shrubs and $\overline{X}_{n'}$ is the mean volume per shrub of the $n' = 569$ shrubs in the eight plots used. The variance is estimated by multiplying the average biomass per shrub $\overline{Y}_{\text{ds}}$ by the average number of shrubs per square meter $\bar{Z}$.

$$\overline{\bar{Y}} \cong \overline{Y}_{\text{ds}}\overline{Z} \tag{8.12}$$

Its variance is estimated by

$$\text{var}\left(\bar{Y}\right) \cong \overline{Z}^2 \text{ var}\left(\overline{Y}_{\text{ds}}\right) + \overline{Y}_{\text{ds}}^2 \text{ var}\left(\overline{Z}\right) \tag{8.13}$$

The optimum ratio of n' to n can be obtained from Equation (8.4) and rearranged to obtain

$$\frac{n'}{n} = \sqrt{\frac{\left(S_y^2 - S_{yx}^2/C_{n'}\right)}{S_{yx}^2/C_n}} \tag{8.14}$$

The highest correlations were obtained with volume (length × width × height) and length, which were the variables chosen for the regressions.

A cost ratio of C_n to $C_{n'}$ is 120:1. That is, it was 120 times as expensive to clip as to measure dimensions, and the total cost was \$1000. This resulted in optimum sample sizes of $n = 7$ and $n' = 184$. It was observed that different categories of plant parts had different optimum allocations.

Bryant and Kothmann (1979) studied the variability in predicting edible crown leaf browse on 12 different species in western Texas. Biomass estimates were made with regression techniques using crown volume–weight relationships. They concluded that a log–log function may yield the best results for large species. That is

$$\text{log weight} = a + b \text{ log (crown volume)} \tag{8.15}$$

where a and b are the usual linear regression model coefficients. Small species may require a quadratic function for best results.

The quadratic form is

$$\text{log weight} = a + b(cv) + c(cv)^2 \tag{8.16}$$

or

$$\text{log weight} = a + b(cv)^2 \tag{8.17}$$

where cv is crown volume and the coefficients are as defined previously. Species with an irregular form and very little foliage may require other mathematical relationships.

Stem diameter

The relationship of total above-ground weight and leaf weight to basal stem diameters was determined by Brown (1976) for 25 northern Rocky Mountain shrub species. The highest correlations were obtained for the natural logarithm of basal diameter and total above-ground weight and leaf weight for all species. Addition of the natural logarithm of stem length did not increase precision greatly, even though its contribution in accounting for biomass variation was significant. Regression components of selected species are given in Table 8.3. Note that the overall variation increases and that the log (basal diameter) does not account for as much of the weight variation (measured by R^2). This decrease occurs because the relationship between log weight and log stem basal diameter is not consistent for all species. Brown recommended that geometric means of diameter ranges be used for both low- and high-diameter classes, while medium-diameter classes could be converted to either an intermediate value or a geometric mean.

Table 8.3 Regression components for estimating leaf and total above-ground weights of sample woody shrub species and three shrub groups using linear equation

$$\ln (\text{weight, g}) = a + b \ln (\text{basal diameter, cm})$$

(From Brown J. K. 1976. Estimating shrub biomass from basal stem diameters. *Can. J. Forestry Res.* **6**: 154–158.)

Species	n	Range of sample diameters (cm)	Leaf weight			Total above-ground weight		
			a	b	r^2	a	b	r^2
		Low shrub						
Symphoricarpos albus (L.) Blake (snowberry)	31	0.2–1.2	1.848	1.721	.68	3.490	2.285	.88
Vaccinium globulare Rydb. (blue huckleberry)	44	0.3–1.7	1.480	2.537	.84	3.388	3.150	.97
Combined species (5)	295	0.2–1.7	2.033	2.165	.67	3.565	2.667	.91
		Medium shrub						
Artemisia tridentata Nutt. (big sagebrush)	22	0.8–6.9	1.603	1.888	.75	3.161	2.242	.92
Juniperus communis L. (common juniper)	23	0.8–2.9	3.414	1.650	.80	4.081	2.202	.92
Combined species (5)	226	0.3–6.9	1.945	2.363	.66	3.580	2.853	.92
		High shrub						
Amelanchier alnifolia Nutt. (serviceberry)	39	0.4–4.5	1.691	2.111	.83	3.607	2.887	.99
Acer glabrum Torr. (mountain maple)	31	0.4–3.7	1.868	2.038	.89	3.634	2.752	.98
Combined species (4)	222	0.4–6.3	1.930	1.974	.79	3.507	2.697	.95

Twig measurements

It is usually more convenient to predict browse production from twig length, which can be measured accurately and quickly, compared to some other methods. Branch dimensions were used by Whittaker (1965) to estimate branch production of woody species in the Great Smoky Mountains of the eastern United States. Shrub and tree biomass production were predicted by counting and measuring branches from stem apex down and data were recorded for a number of variables. A matrix of mean coefficients of correlation among variables shows high correlations with many variables (Table 8.4). However, the prediction of branch weight for evergreen and deciduous shrubs should use branch diameter and functions of branch diameters, as indicated by coefficients listed in column 2 of Table 8.4.

Correlations were found to be useful for the estimation of dry weights of branch wood with bark, current twigs and leaves, and branch production from other measurements. The highest correlation for branch weight (wood with bark) was with powers of the diameter. Current twig weight (with leaves) was strongly correlated with the number of twigs and with the branch dry-weight–age relation.

Ratios of twig length to twig weight were recorded by Halls and Harlow (1971) at 1 year intervals over a 5 year period in Texas and Virginia. Total length was the single best variable for predicting weight ($r = 0.977$). However, ratios of twig weight to length were inconsistent among years and locations. Basile and Hutchings (1966) determined the twig diameter–length–weight relations of bitterbrush (*Purshia tridentata*) in Idaho. Regression analyses were conducted for biomass

Table 8.4 Coefficients of correlation for branches of shrubs, Great Smoky Mountains[a]. (Republished with permission of Ecological Society of America, from Branch dimensions and estimation of branch production. Whittaker, R. H. 1965. *Ecology* **46**: 366; permission conveyed through Copyright Clearance Center, Inc.)

	1	2	3	4	5	6	7	8	9
Means	.50	.71	.74	.74	.70	.71	.73	.69	.70
1. Age		.58	.51	.61	.70	.62	57	.52	.56
2. Branch weight	.52		.91	.82	.77	.72	.88	.90	.93
3. Weight/age	.46	.92		.80	.74	.70	.87	.89	.91
4. Diameter	.60	.82	.76		.81	.97	.97	.91	.88
5. Length	.61	.75	.72	.81		.80	.78	.72	.82
6. Log diameter	.60	.69	.66	.95	.82		.89	.80	.77
7. 4^2	.54	.88	.80	.96	.74	.84		.98	.95
8. 4^2	.47	.86	.78	.88	.64	.72	.98		.97
9. $4^2 \times 5$	.52	.93	.84	.88	.79	.73	.95	.94	
Means	.53	.70	.75	.74	.73	.72	.70	.64	.68

[a]Upper-right diagonal matrix, mean coefficients for branch samples from evergreen shrubs; lower-left diagonal matrix, for branch samples from deciduous shrubs. Top and bottom lines are means of correlation for one variable with all other variables for evergreen shrubs at the Top and deciduous at the bottom.

weight W on stem diameter D, length L, and diameter + length ($D + L$). Twig weight was highly correlated with length ($r = 0.86$, where $W = -0.063 + 0.057L$) and with diameter + length ($r = 0.95$, where $W = + 4.56D + 0.0301L$). Provenza and Urness (1981) studied these relationships for blackbrush (*Coleogyne ramosissima*) in western Utah. In both cases, the correlation between twig diameter, length, and weights was sufficiently consistent to estimate browse production.

Ferguson and Marsden (1977) also used twig diameter, length, and weight relations to estimate browse production of bitterbrush. Thirty shrubs were randomly selected, and each shrub canopy was visually divided into lower and upper halves and into north and south halves, thus quartering the canopy. Three unbranched twigs taken over a wide variety of twig lengths were collected from each quarter. Length and diameter were measured. Mean diameter was calculated from two measurements taken perpendicular to each other. Regression analysis showed a high correlation ($r = 0.752$ to 0.807) of length to diameter and weight to diameter squared ($r = .893$ to 0.929).

Twig and stem measurements

Twig and stem measurements for estimating browse were used by Schuster (1965) for eight species in southern Texas. Shoot weights (twigs and leaves) and twigs (without leaves) represented summer and winter browse, respectively. Measurements recorded were main stem diameter at groundline, number of live twigs per plant, total length of twigs, and total weight of twigs with and without leaves. The best prediction of biomass production ($r = 0.968$) was a combination of twig numbers and lengths. The relationship was, however, not consistent among species. Twig length was the single variable most closely correlated with both shoot and twig weight for most species. Twig counts had the lowest correlations with weight. Stem diameter squared was also highly correlated to weight. The twig count method to measure biomass of hardwood browse for deer was used by Shafer (1963) in Pennsylvania (USA). Shafer compared the twig count method with weight estimates and the clip-and-weigh method (Figure 8.3). Thirty circular plots, 9.2 m^2, were set up to include all browse between ground level and 1.8 m. Ocular estimates of green weight of browse, total and by species, within a plot were made and browse was clipped. Results of the clip-and-weigh method showed no significant differences among the three methods. There were differences, however, in the average time required per plot for each method. Results indicated that the twig count method was almost as rapid as the weight estimate and just as accurate as the clip-and-weigh method.

Bartolome and Kosco (1982) described an "architectural model" for estimation of browse productivity of deerbrush (*Ceanothus integerrimus*) and other shrubs. Their model describes the stem and branch architecture in terms of first-, second-, third- and fourth-order branching. The primary stem that forms the main supporting

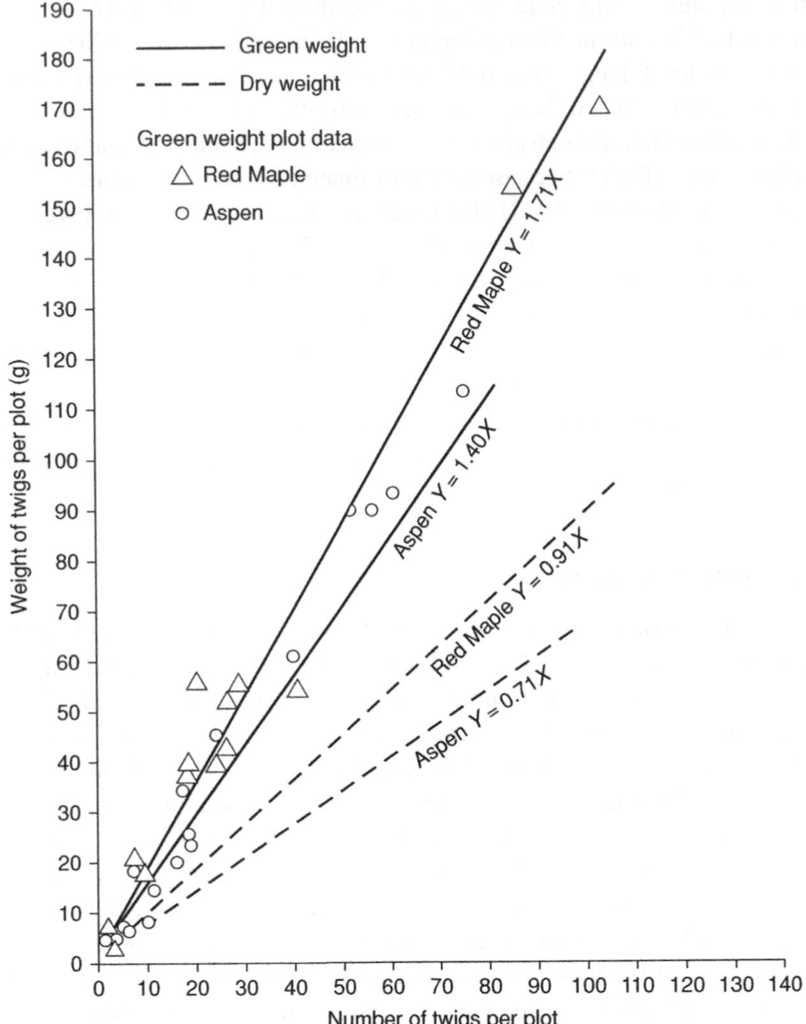

Figure 8.3 Results of the clip-and-weight method, showing the relationship of weight of twigs per plot to number of twigs per plot for red maple and red oak. (Redrawn from Shafer E. L. 1963. The twig-count method for measuring hardwood deer browse. *J. Wildl. Manage.* **27**: 428–437.)

structure is first order. The lateral branches arising from the main stem is second order, and so on. Basal diameter of the branches (second order) arising from the primary stem predicted leaf and branch weights with $r^2 = 0.97$ using an allometric transformation in a linear regression. The model is of the form

$$Y = ax^b \tag{8.18}$$

or

$$\ln y = a + b \ln X \tag{8.19}$$

where y and x are branch weight and diameter, respectively, within a given order of branching.

Growth rings

Width of growth rings has been used in some cases to determine average annual productivity of shrubs. Ring width differences within a site normally do not vary much, but site-to-site differences may be great. Some woody plants, however, do not have well-defined concentric growth rings. Davis *et al.* (1972) reported 20% difference within site and as much as 600% among sites for width of rings. Younger plants have larger growth rings as compared to older plants. This may result in an overestimation of production. To eliminate the problem caused by younger plants with larger rings, individual ring widths are standardized by dividing each ring width by average ring width of that particular plant. The growth ring widths are then expressed as a percentage of the average width. Production of a site should then be expressed as a percentage of the site's average annual production. The stem or branch cross-sections may be stained with iodic green or sanded and polished to measure growth ring widths.

Xylem rings usually increase at three heights of the plant: base of the shoot, 2 cm above surface of forest litter, and in the middle and upper 2 year-old part of the shoot, as measured by Moszynska (1970). Production can be estimated by multiplying the ratio of area of the current year's xylem ring to the area of cross-section of the shoot by the biomass of old shoots

$$W = \frac{D^2 - d^2}{D^2} \tag{8.20}$$

where D is the diameter of the shoot and d is the diameter corresponding to growth in previous years. Davis *et al.* (1972) also estimated browse production from shrub ring widths in Nevada, USA. Productivity was determined by clipping the current year's growth on twenty 0.88 m^2 plots. Stems were measured to the nearest 0.01 mm with a Craighead–Douglas dendrochronograph. Ring widths were measured along two radii on each cross-section and averaged for each year. Plants measured were winterfat, shadscale, bud sagebrush, spiney hopsage, and big sagebrush. The factors were analyzed by all possible correlations followed by stepwise regression with production as the dependent variable. Many of the factors correlated significantly with production, but widths of growth rings of shrubs accounted for most of the variation in production. The correlation coefficient r of production with the width of growth rings ranged from 0.615 for bud sagebrush to 0.972 for big sagebrush. Big sagebrush and shad-scale ring widths varied exponentially with production, while a linear relationship expressed production from ring widths for the other shrubs.

8.2.2 Considerations of shrub biomass sampling

Identifying current growth

An important concern in the measurement of browse production is the definition of what plant parts constitute usable and available forage. Therefore, many field technicians tend to develop their own methods of measurement standards that result in a lack of a common basis for data. It is difficult to objectively define usable and available browse because animal species preference, season, and grazing pressure all determine the amount of utilization. Availability of forage is influenced by terrain, season, density of vegetation, cover, height of browse from the ground, and growth form of plants. Brown (1954) suggested that production of shrubs should be based on the portion of the current year's twig growth that is edible and available to animals. Quite often, bark of new growth is either different in color from that of the old growth or covered by hairs. In contrast, color of old growth appears grayish in color and the epidermal texture is different when compared to new growth. In some species, leaves occur on the current year's twigs only. However, this is not true in the case of evergreen species. A ring of bud-scale scars on some species or a slight swelling of the stem on other species may denote the point of origin of current growth. In actual practice, the ability to distinguish accurately between old and new plant growth is largely a matter of experience and familiarity with the growth characteristics of the species.

Gysel (1957) used twig diameter to estimate forage biomass, but confined measurements of current growth to twigs 3 mm or less in diameter. Forage up to 1.8 m above ground was considered available to deer. In a mixed oak-pine stand in Virginia, Barrett and Guthrie (1969) considered browse to be current annual terminal growth of all woody species from the ground to a height of 1.5 m. A standard range survey procedure has been to consider current annual twigs within 0.8–1.5 m of height above ground level available as forage. However, densely clumped growth of woody plants, and dense short twigs induced by close browsing, pose problems in the determination of forage available.

Wight (1967) studied the effect of plot size on yield estimates of Nuttal saltbrush (*Atriplex nuttallii*) in Wyoming, USA. An 18 × 18 m plot of the saltbrush was completely harvested using 0.3 × 0.3 m sampling units. Contiguous 0.09 m^2 plots were combined into larger units of various sizes and shapes. Wight found that the size of the sampling unit had a pronounced effect on estimated yield variation. Increasing the sampling unit from 0.3 to 5.5 m^2 caused the coefficient of variation ($100s/\bar{x}$) to decrease from 134 to about 30%, while enlargement beyond the 5.5 m^2 size had little effect on the coefficient of variation. The degree of aggregation of the saltbrush was responsible for the lower variation with the larger plot size.

Efficiencies in estimating shrub yields can be gained by using long narrow plots when shrub individuals are clumped or aggregated. Then, more variability in biomass is included within each plot compared to that found in square plots. Wight (1967) found that efficiency of rectangular plots decreased when the smaller dimension of the plot exceeded 1.5 m. Therefore, spatial distribution of shrubs and stems per shrub must be known to select appropriate plot size and shape. In all cases of sampling, whether by twig reference units or from quadrats, relative coordinates for location of individual sampling units should be recorded so that spatial effects can be accounted for and measured, and variances adjusted accordingly within each study area.

Regression equations

Regression equations for prediction of browse production from plant characteristics such as length and diameter of twigs would be more useful if these could be applied equally well throughout a large geographic area. However, this is not generally so because the magnitude of differences among years, seasons, and locations is usually great. Therefore, individual site–year equations should be used until such time that sufficient data are available for evaluation of site–regional equations.

A representative sample of the current year's twig growth must be obtained from the chosen site through a random sampling procedure to develop a regression equation for predicting browse production. Clip the twigs cleanly at the base and avoid browsed or broken twigs. Include twigs from all sides of the shrubs and from lower as well as upper branches. Obtain a range of twig lengths from minimum to maximum length. The clipped twigs should be placed in plastic bags to prevent drying out until weighing.

8.3 Forest biomass

Tree, and subsequently, forest biomass may be a more important measure to mankind than volume of tree boles since all biomass materials are being used in much of the world. Yet, little information has been collected on other biomass components. Obtaining tree biomass estimates from correlated stem and canopy measurements follows the same procedures as those used for shrubs. Mensuration books specifically deal with volumetric estimates for timber saw-logs, and so forth. On the other hand, biomass estimates of stems, leaves, and bark often require more detailed measurements to adequately predict the standing biomass of trees. Most procedures use statistical models that are solved by regression techniques. Detailed procedures are provided by Avery and Burhart (2003), Husch *et al.* (2003), West (2004), van Laar and Akca (2007) and Kalkhan (2011).

8.3.1 Regression models of biomass estimation

Plant ecologists should use statistical models to obtain estimates of tree biomass on an individual basis and sum over trees to obtain stand estimates. Components of biomass (stems, leaves, etc.) for trees can be estimated on an individual basis for stands. Stem diameter and tree height are easily measured, but secondary branches can present problems. The use of dbh, as a single measure, is often considered as sufficient to estimate the biomass of the tree (Baskerville 1965). Therefore, time is saved by this easily obtained single measure. For some plant ecologists, both tree height and dbh seem to be sufficient, but others want the geometry of the canopy measured for a more complete biomass description. The references listed above (Section 8.3) should be consulted. Models used are simple and commonly include:

$$\text{Allometric} \quad y = ax^b \tag{8.21}$$

$$\text{Exponential} \quad y = ae^{bx} \tag{8.22}$$

$$\text{Quadratic} \quad y = a + bx + cx^2 \tag{8.23}$$

where y is the biomass component to be estimated, x is diameter and a, b, and c are regression coefficients. Husch *et al.* (2003), West (2004) and others discuss these models more fully. The allometric Equation (8.21) has been the most reliable in many studies.

Ouellet (1985) used a non-linear equation to predict individual total tree biomass for several species and for various components of individual trees: stem wood, wood and bark of stem, and crown biomass (leaves and secondary branches). The model:

$$Y = B_1 D^{B_2} H^{B_3} + e \tag{8.24}$$

can be used to predict the biomass of a given component Y, where B_i ($i = 1, 2, 3$) are constants estimated by non-linear regression methods, and e is the error term (West 2004). The model gave reasonably good results in predicting most component biomass for several species, as indicated by the percentage of variation accounted for (r^2) (Table 8.5).

A number of authors have reported results of predicting tree biomass by regressing biomass against various tree dimensions. Bole diameter, tree height, depth of crown, and crown diameters are commonly used in regression equations. To estimate the parameters of equations, individual trees must be dissected, measured, and the biomass determined by component of interest. In general, the most successful

Table 8.5 Coefficients of determination R^2 for predicting oven-dry biomass (kg) of total above-ground tree and components. (From Ouellet D. 1985. Biomass equations for six commercial tree species in Quebec. *Forestry Chron.* **61**: 218–222.)

Species	Total	Stem	Crown	MS[a]	MS wood	MS bark
Eastern white cedar (*Thuja occidentalis* L.)	.96	.98	.70	.97	.96	.94
Eastern hemlock (*Tsuga candensis* L. Carr.)	.98	.98	.88	.98	.97	.93
Red maple (*Acer rubra*)	.98	–	.82	.98	.98	.95
Beech (*Fagus grandifolia* Ehr.)	.98	–	.80	.98	.97	.91
Black ash (*Fraxinus nigra* Marsh.)	.97	–	.45	.97	.97	.95
White ash (*Fraxinus americana* L.)	.98	–	.79	.99	.99	.94

[a]Total merchantable stem (MS) to 9 cm at top for conifers; to 9 cm in branches for deciduous species.

equation is non-linear in form, of the type given by Equation (8.21) or some variant in a log-transform model. If coefficients (parameters) are found to be numerically close in value for equations of several species, then a single equation can be used to predict biomass for several species. That is, by using measurements taken of the predictor values(s) from individual species, the equation to predict biomass can be solved.

Many published equations can be used to derive preliminary estimates for components of the tree biomass. However, such equations are usually specific to both species and habitat characteristics and may not be useful if a high degree of precision is required. The use of regression equations to predict biomass of trees, like those of shrubs and herbaceous plants, are usually data specific. That is, equations are useful only within the range of data used to estimate model coefficients, and models should not be used to extrapolate predicted values outside of data used to formulate the model. In summary, the best overall estimate of tree biomass component or total may be a logarithmic function of dbh. For example, log(biomass) = $a + b \log(dbh)$.

8.3.2 Bark biomass

Tree bark has a number of commercial uses, and bark production may be an important measurement to be made, along with other tree biomass components. It is often treated separately because equations usually are formulated to predict volume of bark rather than biomass. Detailed descriptions of estimation of bark parameters are provided by Husch *et al.* (2003).

8.4 Tree foliage biomass

Measurements of tree foliage biomass are needed to determine their contribution to the total productivity of a forest community. Foliage production can be used as a measure of evaluating site potential for forests because foliage biomass is determined by stand and site characteristics. In arid and semi-arid regions of the world, trees are lopped for feeding foliage to animals, and materials are used as fuel for cooking. In addition, medium-sized trees, young trees, and coppice shoots are browsed by livestock, as well as by large wildlife herbivores. Even large trees are within browsing range of camels and giraffes.

Production of foliage or browse of young trees and coppice shoots can be estimated by the direct harvest or non-destructive and indirect methods described earlier for measuring shrub production. However, in the case of tall trees, harvesting is difficult. Therefore, the most feasible method to estimate production of tree foliage is through indirect methods. The ecologist should find some easily measured tree characteristics that are correlated with the biomass of leaves, stems, and branches. Crown diameter, foliage cover, basal area, diameter at breast height, diameter at the base of live crown and/or branches, tree height, height of crown alone or in combination, have been found to be reliable predictors of foliage biomass.

Reference unit technique

The reference or sample weight unit technique is used to estimate leaf biomass. Trees are first classified into groups based on compactness of the canopy foliage. As an example, Mason and Hutchings (1967) used three groups for Utah juniper (*Juniper utahensis*) (sparse, medium, and dense). A sample weight unit with the average foliage and fruit is selected and used as a standard for estimating fruit and foliage yields in each group. The number of weight units on each tree is then counted. Foliage and fruit are clipped from a sample unit, air-dried, and weighed. The total weight of the foliage and fruit on each tree is then estimated by multiplying the number of weight units by the weight of the sample unit. Estimates of mean foliage and fruit production for Utah juniper were predicted within 10% of the mean with 95% confidence for samples of 20 trees within each crown class. Each species would have to be studied for the appropriate unit used as a reference for the current year's growth. Sample size, needed to obtain sample adequacy, employs the standard sample size equation.

Crown diameter

Tree foliage yield can be estimated from crown measurements. The correlation between crown and foliage production can be evaluated more effectively by inclusion of foliage denseness and site factor measurements. Tree heights and crown diameters are measured to the nearest 0.15 m for each tree, and canopies of trees are classified into groups; for example, sparse, medium, and dense, based on the compactness of foliage. Mason and Hutchings (1967) estimated foliage production of Utah juniper from measurements of crowns on 400 m^2 randomly located plots. Logarithmic equations provided the best prediction equations. Correlation coefficients ranged from 0.88 to 0.98 for these equations relating foliage and fruit production to crown diameter. The equations, however, differed in coefficient values from site to site. Relations of current foliage yield of juniper to crown diameter for three foliage classes on five range sites were shown to follow the same equation form, an exponential.

These equations relating weight of foliage and fruit to various tree measurements were developed by relating:

- Logarithm of yield with logarithm of crown diameter
- Logarithm of yield with logarithm of height
- Logarithm of yield with logarithm of crown diameter and logarithm of height
- Yield with height squared
- Yield with height
- Yield with crown diameter
- Yield with crown diameter squared Yield with crown surface
- Yield with crown volume.

Equations differed among sites so the site effect should be added to the model. Most investigators ignore such effects by developing an equation for each site. The use of a site term is easily incorporated and, in such cases, uses only one equation. For instance, the site effect is z in Equation (8.7).

Crown cover and basal area

Some studies have indicated that a high degree of correlation exists between leaf biomass and stem area at ground level or at breast height (1.4 m) and canopy area. Hutchings and Mason (1970) estimated foliage yields of gambel oak (*Quercus gambellii*) from foliage cover and basal area. Five to ten 20 m^2 square plots were used to sample dense stands of oak. Scattered stands were sampled by ten 40 m^2 plots. The number of stems per clump was recorded and foliage over 1.8 m in height was recorded separately. Sample weight units were clipped for foliage, acorns, and annual twig growth. Total air-dry weight, basal area, and foliage area were computed as follows

$$\text{Total air-dry weight} = \text{air-dry weight of samples} \times \text{number of weight units in each clump}$$

$$\text{Basal area} = \text{average stem diameter}^2 \times \text{number of stems}$$

$$\text{Foliage area} = \text{width of segment} \times \text{length of segment}$$

On most sites, foliage area was associated more closely with foliage production than either foliage volume or basal area. A multiple regression of foliage area and basal area provided the best prediction equation and accounted for over 75% of variation.

Stem and/or branch measurements

Periodic annual growth of stemwood, tree diameter, branch diameter, and circumference have also been found to have a high correlation with foliage growth and/or total biomass. The reason for this correlation is that the amount of increment of stemwood or branchwood is a function of the amount of foliage that is photosynthesizing in a given period of time. Measurements of all first-order branches on sample trees, or on all trees on sample plots, can be used to estimate foliage biomass.

Several studies have found stem or branch diameter at the base of live crown as a better predictor of foliage biomass. All these measurements, except periodic annual growth, are easy to obtain. Kittredge (1944) found a well-defined linear trend when he plotted dry weight of jackpine (*Pinus spp.*) needles over the 5-year periodic annual increment in cubic feet. He also found a linear trend when leaf weight was plotted against stem diameter on double logarithmic paper. Rodionov (1959) obtained correlation coefficients of 0.90 to 0.93 between diameter at root collar and fresh weight of foliage for 2- and 4-year-old shelterbelts of oak (*Quercus spp.*) and ash (*Fraxinus spp.*) trees in Russia. He also found correlation coefficients of 0.89 and 0.86 for foliage weight with branch diameters for poplar (*Populus spp.*) and willow (*Salix spp.*), respectively. Elkington and Jones (1974) estimated fresh weights of leaves from regression of branch circumference on fresh weight of leaves per branch from

$$Y = 11.93 + 1.629X \tag{8.25}$$

where Y represents leaf weight per branch and X represents branch circumferences. Measurements included several species of trees in Greenland, and they concluded that their model did not apply to other areas.

Elkington and Jones also suggested equations for predicting branch dry weight BW, leaf dry weight LW, and caudex dry weights CW based on a function of average branch circumference BC. In general form

$$\mathrm{BW}^{1/3} = a_1 + b_1\,(\mathrm{BC}) \tag{8.26}$$

$$\mathrm{LW}^{1/3} = a_2 + b_2\,(\mathrm{BC}) \tag{8.27}$$

$$\mathrm{CW}^{1/3} = a_3 + b_3\,(\text{sum of BC}) \tag{8.28}$$

Note that the relationship is one of the cubic root of biomass and that the three biomass classes are linearly related to the branch circumference. Baskerville (1965) found, in balsam-fir–white-spruce–white-birch (*Abies–Picea–Betula*) stands in New Brunswick, that total dry weight of foliage and branches of each species was closely correlated to stem dbh ($r = 0.96$ to 0.99). This early study showed justification for using dbh alone to predict tree biomass. This use of dbh is the result of the bole's growth characteristics producing wood strength throughout to support the entire tree biomass above the dbh. In other words, the tensile properties of the tree bole diameter determine the amount of biomass that can be supported.

Felker *et al.* (1982) investigated stem diameter and height relationships to fresh and dry-weight biomass of tree legumes of *Prosposis*, *Cercidium*, *Olneya*, *Leucaena*, and *Parkinsonia* genera. They computed the following relationship to predict the biomass of the legume tree genera under study:

$$Y = 0.363X - 0.537 \tag{8.29}$$

where Y is the weight in kilograms per tree and X is the stem diameter. Obviously, for trees with diameters less than 1.5 cm, this relationship gives negative biomass. Yet log–log regression of basal stem diameter against biomass was solved for trees smaller than 1.5 cm in diameter.

There has not been much change in statistical sampling nor in the measurement of tree biomass over the past 50 years. There has been much work, however, on the use of remote sensing, both aerial and satellite imagery, in vegetation measurements of cover and biomass in particular. The preponderance of advancements has been made for forested communities and on stands within them. Species are grouped into life forms or other categories for measures, not measured on an individual basis.

8.5 Considerations for tree biomass sampling

8.5.1 Annual production of tree foliage

Annual production of foliage has been estimated from values of both standing biomass and litter fall. Estimates from standing biomass usually assume that annual production is equal to the weight of 1-year-old leaves and their "stems" on trees at the end of the growing season. This assumption overlooks any decline in weight of individual leaves at the end of the growing season. These losses are due to insect feeding and year-to-year fluctuations in foliage production. Hence, underestimates of total production are obtained. Foliage production estimates from litter traps involve similar assumptions and are also dependent on the type of litter trap employed. Whittaker (1966) based leaf production of deciduous species directly on foliage biomass and in evergreen trees, on a comparison of two estimates: foliage biomass (from regression on dbh) divided by years of leaf persistence, and from foliage production ratios to stem wood production. No correction was made for leaf consumption by animals, but an increase of the estimates of net production should be made to allow for unsampled plant parts such as bud-scales, flower parts, cones, and other fruits.

8.5.2 Production of tree biomass

Above-ground woody biomasses can be estimated by two independent approaches. One estimate is to sum stem wood and bark weight (Whittaker 1965) and normal foliage weights of deciduous and needleleaf evergreen forest (Whittaker 1962). The other estimate is obtained from total tree weights, and in some species branch and foliage weights, which are predicted from dbh values using regression equations already published for a species or genera.

Production estimates for tree strata can be summarized, as based on the work by Whittaker (1966):

1. Stem wood growth is estimated by multiplying estimated volume increments by wood density values.
2. Stem production is estimated from the relation: $\Delta S/S = k(\Delta A/A)$, where S and ΔS are stem biomass and production, respectively, A and ΔA are basal area and basal area increment for a tree, and k is a constant.
3. Stem bark growth estimates are obtained by multiplying the stem wood growth estimate by the ratio of bark to stem wood basal area and multiplying this by an arbitrary correction for difference in relative growth rate.
4. Branch production is estimated from the equation $\Delta B/B = k(\Delta S/S)$, where ΔB and B are branch production and weight, and ΔS and S are wood production and weight. The separate estimates of branch, stem wood, and stem bark production should be checked against direct estimates of woody shoot production from the biomass and basal area increment ratio.
5. Leaf production of deciduous species is obtained directly from foliage biomass. Foliage production in evergreen species is based on two estimates: foliage

biomass (from regression in dbh) divided by years of leaf persistence, and foliage production from ratios to stem wood production.

6. Leaf, flower, and fruit loss is visually estimated. No correction is made for leaf consumption by animals, but the estimates of net production may be increased by a percentage to allow for unsampled plant parts such as bud-scales, corollas, cones, and other fruits.
7. Net shoot production estimates are based on summing individual estimates for plant parts.
8. A large sample size for almost any sampling unit will either reduce or eliminate bias associated with the unit.

Biomass estimates are more reliable for stands of small trees for which appropriate regressions on stem diameter are available. However, such estimates can be affected by errors when equations are extrapolated to larger-sized trees. Estimates of biomass, made from parabolic volumes, are significantly affected by branch:stem ratios, while other estimates may be affected by actual:estimated volume increment ratios. These variables should be studied carefully before they are used to predict tree biomass.

8.6 Selection of sampling units for tree biomass

Kohl *et al.* (2006) provided detailed information on selection of trees for fixed area sampling units. In the simple random selection process, every member of the population has an equal probability of being selected. On the other hand, in systematic sampling, only the first unit is selected randomly and all subsequent units are equally spaced (Avery and Burkhart 2003). If a vegetation type occurs in scattered clumps in the area to be sampled, then the number of sampling units in each clump should be proportional to the size of the clump, and in the statistical and forest literature this is referred to as "sampling with probability proportional to size" and abbreviated as "PPS sampling."

In simple random or systematic sampling, all the sampling units are of the same size and shape. All trees in the sampling unit are measured for diameter and height classes, and the class of trees is sampled in proportion to the number of trees in that class. In some forests, there are more small trees than large trees, and, thus, smaller trees would be sampled more intensively than bigger trees. Therefore, in uneven-aged forests, a sampling design that is based on the probability of a tree being selected in a sampling unit proportional to the frequency of its size would give more precise estimates (a type of stratified random sample). One type of probability sampling is 3P (probability proportional to prediction) (West 2004). This is a modification of the double-sampling procedure described earlier in this chapter.

In the double-sampling procedure, the variables of interest are guessed (estimated) on all sampling units n. On a sub-set of sample units n', the variables of interest are both guessed and measured. The sub-sample n' is randomly drawn from the larger sample size n. On the other hand, in 3P sampling, the variable of interest, usually tree volume, is guessed on all the trees in a stand. The estimated volume of the tree is compared with a randomly selected integer from a list. If the volume estimate is greater than or equal to the number, the tree is measured for precise volume determination. The list has a set of integers from 1 to a maximum tree volume that can be expected. Further details on 3P or other similar sampling procedures are presented in textbooks on forest mensuration and measurements. Examples are Husch *et al.* (2003) and Kohl *et al.* (2006).

8.6.1 Sampling procedures

Three main sampling procedures are used to determine the biomass of tree components: unit area, the average tree, and regression analysis. The unit area method requires the collection of weights for tree components or foliage on a sample plot area or series of sample plots. The average tree method requires selection of trees considered to be of average biomass based on knowledge of the linear dimensions of the trees (West 2004). Foliage biomass per unit area is then determined by multiplication of the average number of trees in the unit area. This procedure, however, is not applicable in mixed-age forests. It is useful only where a rough estimate of total biomass is required. Baskerville (1965) reported that the average tree method could lead to gross errors in biomass estimation. The regression analysis method uses a mathematical relationship between weights of tree components and one or more of the easily measured tree dimensions. Regression equations may not be appropriate if sampling is restricted to a portion of population such as small trees or large trees. Results can be improved by stratifying the population into 5 or 10 size classes and selecting trees that are nearest to the average of each class.

The sampling error of random, stratified, and strip sampling designs was studied by Hasel (1938), West (2004), and Kohl *et al.* (2006). Stratified sampling was more efficient than random sampling. The strips were used because of clustering within strata. Systematic sampling within homogeneous strip plots was more efficient than random sampling. Random sampling and, to a greater extent, stratified random sampling of trees usually will yield residual mean squares about the regression line that are generally smaller than those found using all trees. Estimated standing weights for tree biomass from stratified sampling may be skewed by a few large overestimates. Stratification will reduce the number of trees required for a given precision. Variances and confidence intervals may be calculated using the methods of Cochran (1977) and Thompson (1992). Madgwick and Satoo (1975) found that stratified random sampling gave smaller residual mean squares about regression than those found for random sampling or using all trees in plots.

Table 8.6 Relative efficiency of plots of varying size and shape. (From Hasel A. A. 1938. Sampling error in timber surveys. *J. Agric. Res.* **57**: 713–736.)

Plot size and shape	Mean square	Efficiency (%)	Relative size of sample for a given precision of estimate
1 ha: 50 × 200 m	282.4961	100.00	1.00
2 ha: 100 × 200 m	449.3926	62.86	1.59
4 ha: 50 × 800 m	566.0602	49.90	2.00
100 × 400 m	711.8291	39.68	2.52
200 × 200 m	687.4958	41.09	2.43

Hasel (1938) studied relative efficiency of plots of varying sizes and shapes. The results of random sampling for size and shape of plot are given in Table 8.6. Among 4-ha plots, the long and narrow shape was most efficient. Variations in forests are usually greatest at right angles to contour lines, and sampling at right angles to the contours with long, narrow plots is more advantageous.

Bormann (1953) conducted the classic study on effective size and shape of plots in the Piedmont plateau of North Carolina. An area of 140 m on a side was divided into 4900 plots of 2 × 2 m. Basal area was determined for two distributions: (1) one containing 82.5% of the total basal area and represented the vegetation type and (2) one containing only 11% of the total basal area sporadically distributed over the area. Thus, each plot had a value for each distribution associated with it. Plots were combined in various sizes, shapes, and orientations, and the variances were calculated. Long narrow plots for the first distribution were most efficient when the main axis ran across contours or soil banding. Variances decreased with plot length, but the smallest variances were obtained with wider plots. There was little difference in variance for different plot orientations. Variances increased with length if plots paralleled the contours in the second distribution but decreased if the orientation was across contours. That is, an increased edge-effects error for long narrow plots. Longer plots were considered easier to randomly locate in the field than were shorter plots.

The following four designs would be best for the first kind of distribution: 4 × 140 m (B), 10 × 70 m (A), 10 × 140 m (B), and 10 × 140 m (A). The second distribution was best sampled by 4 × 140 m (B) and 10 × 140 m (B) plots. Type A assumes that the long axis is parallel to the direction of greatest variation, and type B is perpendicular to that variation.

Plot Size and Shape

The basic considerations in selecting an optimum plot size and shape are the same as discussed in Section 8.1.5 for measurement of herbaceous vegetation. In forest inventory, the sampling units are either in the shapes of strips or circular plots. The strips are usually 20 m wide and run from one edge of the study boundary to the other. The wide strip is reduced to 10 m in thick growth and increased to 40 m if the trees are widely scattered. The circular plots are usually 11–12 ha in North America and 0.01–0.05 ha in Europe. Although circular plots are more common, some authors have used square and rectangular plots.

The only advantage of strip sampling is that time is saved in traveling from plot to plot. However, an optimum size of plot is one in which equal amounts of time are spent on travel and on plot measurement (Zeide 1980). Husch *et al.* (2003) provide further details on optimum size and shape of plots for tree measurements.

8.7 Bibliography

Andariese S. W. and Covington W. W. 1986. Biomass estimation for four common grass species in northern Arizona ponderosa pine. *J. Range Manage.* **39**: 472–473.

Anderson D. M. and Kothmann M. M. 1982. A two-step sampling technique for estimating standing crop of herbaceous vegetation. *J. Range Manage.* **35**: 675–677.

Andrew M. H., Noble I. R. and Lange R. T. 1979. A non-destructive method for estimating the weight of forage on shrubs. *Aust. Range J.* **1**: 225–231.

Andrew M. H., Noble I. R., Lange R. T., and Johnson A. W. 1981. The measurement of shrub forage weight: Three methods compared. *Aust. Range J.* **3**: 74–82.

Avery T. E. and Burhart H. E. 2003. *Forest Measurements*, 5th edn. McGraw-Hill, Inc.: New York.

Barrett J. P. and Guthrie W. A. 1969. Optimum plot sampling in estimating browse. *J. Wildl. Manage.* **33**: 399–403.

Bartolome, J. W. and Kosco B. H. 1982. Estimating browse production by deer-shrub (*Ceonothus integerrimus*). *J. Range Manage.* **35**: 671–672.

Basile J. V. and Hutchings S. S. 1966. Twig diameter–length–weight relations of bitterbrush. *J. Range Manage.* **19**: 34–38.

Baskerville G. L. 1965. Estimation of dry weight of tree components and total standing crop in conifer stands. *Ecology* **46**: 867–869.

Baskerville G. L. 1972. Use of logarithmic regression in the estimation of plant biomass. *Can. J. Forestry* **2**: 49–53.

Bormann F. H. 1953. The statistical efficiency of sample plot size and shape in forest ecology. *Ecology* **34**: 474–487.

Boutton, T. W. and Tieszen L. L. 1983. Estimation of plant biomass by spectral reflectance in an East African grassland. *J. Range Manage.* **36**: 213–216.

Brown D. 1954. *Methods of Surveying and Measuring Vegetation.* Bulletin No. 42, Commonwealth Bureau of Pastures and Field Crops: Hurley, Berkshire.

Brown J. K. 1976. Estimating shrub biomass from basal stem diameters. *Can. J. Forestry Res.* **6**: 153–158.

Bryant F. C. and Kothman, M. M. 1979. Variability in predicting edible browse from crown volume. *J. Range Manage.* **32**: 144–146.

Campbell R. S. and Cassady J. T. 1949. Determining forage weight on southern forest ranges. *J. Range Manage.* **2**: 30–32.

Christidis B. G. 1931. The importance of the shape of plots in field experimentation. *J. Agric. Sci.* **21**: 14–37.

Clark I. 1945. Variability in growth characteristics of forage plants on summer range in central Utah. *J. Forestry* **43**: 273–283.

Cochran W. G. 1977. *Sampling Techiques*, 3rd edn. John Wiley & Sons, Inc.: New York.

Crafts E. C. 1938. Height-volume distribution in range grasses. *J. Forestry* **36**: 1182–1185.

Dalkes P. D. 1941. Use and availability of the more common browse plants in the Missouri Ozarks. *Transactions of the Sixth North American Wildlife Conference*, 155–160.

Davis J. B., Tueller P. T., and Bruner A. D. 1972. Estimating forage production from shrub ring widths in Hot Creek Valley, Nevada. *J. Range Manage.* **25**: 398–402.

Elkington T. T. and Jones B. M. G. 1974. Biomass and primary productivity of birch (Betula pubescens S. LAT.) in S.W. Greenland. *J. Ecol.* **62**: 821–830.

Felker P., Clark P. R., Osborn J. F., and Cannell G. H. 1982. Biomass estimation in a young stand of Mesquite (*Prosopis* spp.), Ironwood (*Olneya tesota*), Palo Verde (*Cercidium floridium*), and (*Parkinsonia aculeata*), and Leucaena (*Leucaena leucocephala*). *J. Range Manage.* **35**: 87–89.

Ferguson, R. B. and Marsden M. A. 1977. Estimating overwinter bitterbrush utilization from twig diameter-length-weight relations. *J. Range Manage.* **30**: 231–236.

Gysel L. W. 1957. Effects of silvicultural practices on wildlife food and cover in oak and aspen types in northern Michigan. *J. Forestry* **55**: 803–809.

Halls L. K. and Harlow F. R. 1971. Weight-length relations in flowering dogwood twigs. *J. Range Manage.* **24**: 236–237.

Harniss R. O. and R. B. Murray. 1976. Reducing bias in dry leaf weight estimates of big sagebrush. *J. Range Manage.* **29**: 430–432.

Hasel A. A., 1938. Sampling error in timber surveys. *J. Agric. Res.* **57**: 713–736.

Heady H. F. 1950. Studies on bluebunch wheatgrass in Montana and height-weight relationships of certain range grasses. *Ecolo. Monogr.* **20**: 55–81.

Hickey W. C. Jr. 1961. Relation of selected measurements to weight of crested wheatgrass plants. *J. Range Manage.* **14**: 143–146.

Hughes E. E. 1962. Estimating herbage production using inclined point frame. *J. Range Manage.* **15**: 323–325.

Husch B., Beers, T., and Kershaw J. A. Jr. 2003. *Forest Mensuration*. John Wiley & Sons, Inc.: Hoboken, NJ.

Hutchings S. S. and Mason L. R. 1970. Estimating yields of gambel oak from foliage cover and basal area. *J. Range Manage.* **23**: 430–434.

Kalkhan M. A. 2011. *Spatial Statistics Geospatial Information Modeling and Thematic Mapping*. CRC Press: Boca Raton, FL.

Kelly A. F. 1958. A comparison between two methods of measuring seasonal growth of two strains of *Dactylis glomerata* when grown as spaced plants and in swards. *J. Br. Grassl. Soc.* **13**: 99–105.

Kirmse R. D. and Norton B. E. 1985. Comparison of the reference unit method and dimension analysis methods for two large shrubby species in the Caatinga Woodlands. *J. Range Manage.* **38**: 425–428.

Kittredge J. 1944. Estimation of the amount of foliage of trees and stands. *J. Forestry* **42**: 905–912.

Kohl M., Magnussen S., and Marchetti M. 2006. *Sampling Methods, Remote Sensing and GIS Multiresource Forest Inventory*. Springer-Verlag: Berlin, Heidelberg.

Ludwig J. A., Reynolds J. F., and Whitson P. D. 1975. Size-biomass relationship of several chihuahuan desert shrubs. *Am. Midl. Nat.* **94**: 451–461.

Lyon L. J. 1968. Estimating twig production of serviceberry from crown volumes. *J. Wildl. Manage*. **32**: 115–119.

Madgwick H. A. I. and Satoo T. 1975. On estimating the aboveground weights of tree stands. *Ecology* **56**: 1446–1450.

Mason L. R. and Hutchings S. S. 1967. Estimating foliage yields on Utah juniper from measurements of crown diameter. *J. Range Manage.* **20**: 161–166.

Medin D. E. 1960. Physical site factors influencing annual production of true mountain mahogany (*Cercocarpus montanus*). *Ecology* **41**: 454–460.

Morris M. J. 1959. Some statistical problems in measuring herbage production and utilization. *Symposium on Techniques and Methods of Measuring Understory Vegetation. Tifton, GA, 1958*. Publication of USDA Forest Service, 139–145.

Morrison D. A., Le Brocque A. F., and Clark P. J. 1995. An assessment of some improved techniques for estimating the abundance (frequency) of sedentary organisms. *Vegetatio* **120**: 131–145.

Moszynska B. 1970. Estimation of the green top production of the herb layer in a bog pinewood *Vaccinis uliginosi-pinetum. Ekologia Polska* **18**: 779–803.

Mueller-Dombois D. and Ellenberg H. 1974. *Aims and Methods of Vegetation Ecology*. John Wiley & Sons, Inc.: New York.

Murray R. B. and Jacobson M. Q. 1982. An evaluation of dimension analysis for predicting shrub biomass. *J. Range Manage.* **35**: 451–454.

Ouellet D. 1985. Biomass equations for six commercial tree species in Quebec. *Forestry Chron.* **61**: 218–222.

Pearson R. L. and Miller L. D. 1972a. *Remote Spectral Measurements as a Method for Determining Plant Cover*. US/IBP Grassland Biome Technical Report No. 167, Colorado State University: Fort Collins, CO.

Provenza F. D. and Urness P. J. 1981. Diameter-length-weight relations for blackbrush (Coleogyne ramosissima) branches. *J. Range Manage.* **34**: 215–217.

Rittenhouse L. R. and Sneva F. A. 1977. A technique for estimating big sagebrush production. *J. Range Manage.* **30**: 68–70.

Rodionov M. S. 1959. Determining foliage weight in shelter belts. *Bot. Zuhr. SSSR (USSR)* **44**: 333–337.

Schuster J. L. 1965. Estimating browse from twig and stem measurements. *J. Range Manage.* **18**: 220–222.

Shafer E. L. 1963. The twig-count method for measuring hardwood deer browse. *J. Wildl. Manage.* **27**: 428–437.

Singh J. S., Lauenroth W. K., and Steinhorst R. K. 1975. Review and assessment of various techniques for estimating net aerial primary production in grasslands from harvest data. *Bot. Rev.* **41**: 181–232.

Teare I. D. and Mott G. O. 1965. Estimating forage yield in situ. *Crop Sci.* **5**: 311–313.

Thompson S. K. 1992. *Sampling*. John Wiley & Sons, Inc.: New York.

Tucker C. J. 1980. A critical review of remote sensing and other methods for non-destructive estimation of standing crop biomass. *Grass Forage Sci.* **35**: 177–182.

Uresk D. W., Gilbert R. O, and Rickard W. H. 1977. Sampling big sagebrush for phytomass. *J. Range Manage.* **30**: 311–314.

Van Dyne G. M., Vogel W. G., and Fisser H. G. 1963. Influence of small plot size and shape on range herbage production estimates. *Ecology* **44**: 746–759.

Van Laar A. and Akca A. (2007) *Forest Mensuration*. Springer: Dordrecht, the Netherlands.

Waller S. S., Brown M. A., and Lewis J. K. 1981. Factors involved in estimating green biomass by canopy spectroreflectance measurements. *J. Range Manage.* **34**: 105–108.

Wells K. F. 1971. Measuring vegetation changes of fixed quadrats by vertical ground stereophotography. *J. Range Manage.* **26**: 233–236.

West P. W. (2004). *Trees and Forest Measurement*. Springer–Verlag: Berlin, Heidelberg, New York.

Whittaker R. H. 1962. Net production relations of shrubs in the Great Smoky Mountains. *Ecology* **43**: 357–377.

Whittaker R. H. 1965. Branch dimensions and estimation of branch production. *Ecology* **46**: 365–370.

Whittaker R. H. 1966. Forest dimensions and production in the Great Smoky Mountains. *Ecology* **47**: 103–121.

Wiegert R. G. 1962. The selection of an optimum quadrat size for sampling the standing crop of grasses and forbs. *Ecology* **43**: 125–129.

Wight J. R. 1967. The sampling unit and its effect on saltbush yield estimates. *J. Range Manage.* **20**: 323–325.

Wight J. R. and Neff E. C. 1983. *Soil-Vegetation-Hydrology Studies, Vol. II, User Manual for ERHYM: The Ekalaka Rangeland Hydrology and Yield Model.* USDA, ARS, Agriculture Research Service, ARR-W-29.

Wight J. R., Hansen C. L., and Whitmer D. 1984. Using weather records with a forage production model to forecast range forage production. *J. Range Manage.* **37**: 3–6.

Wimbush, D. J., Barrow M. D., and Costin A. B. 1967. Color stereophotography for measurement of vegetation. *Ecology.* **48**: 150–152.

Wisiol K. 1984. Estimating grazingland yield from commonly available data. *J. Range Manage.* **37**: 471–475.

Yiruhan X.-M. X. and Shiyomi M. (2011). Above ground plant mass and mass available to grazing goats in a mountainous shrubland in subtropical China. *Grassland Science* **57**: 119–126.

Zeide B. 1980. Plot size optimization. *Forest Sci.* **26**: 251–257.

9 Monitoring and evaluation

Monitoring and evaluation of vegetation is important to gain further understanding of the structure and function of plants in their natural ecosystems. A field ecologist who only conducts a vegetation inventory of a study area within an ecosystem will not acquire the understanding needed to develop management strategies for that particular system. This is particularly true in land agencies that employ vegetation ecologists. Over a year's time these specialists spend little time conducting vegetation field studies, analyzing, and reporting results compared to time expended on administrative duties. As a result most problems concerning management of native vegetation systems remain unsolved.

Monitoring and evaluation of vegetation characteristics require a database of soil, topographic, and vegetation resources and, in some cases, regional climatic data. Monitoring and evaluation of data obtained from plant communities, vegetation types, and so forth should be described with respect to a unit within a general classification system. In contrast, many vegetation studies are conducted and reported without being identified with a particular vegetation classification system. These kinds of studies may be assumed to have no known classified vegetation–environmental system. In which case, one could assume that the "individualistic hypothesis" of Gleason (1926) is being followed, or at least the data by species can be analyzed using the methods of Gleason, Curtis (1959), and Whittaker (1951, 1967) among others. These latter methods have been said to "eclipse" the Clementsian approach in America (Shimwell 1971).

If the individualistic concept of a species and its environmental factors are followed for inventory, monitoring, and analysis of results, this approach seems to be more in keeping with the real world behavior of a plant species. One should not ignore that all data collected for an individual species is conditional on the influences that associated species and their environmental factors have on that single species. Therefore, data analysis should include multivariate covariance methods to account for these effects. In essence this means that data (e.g., biomass) will be collected on all species occurring with the single species in a sampling unit. Threatened and endangered plant species should not be excluded from the requirement,

Measurements for Terrestrial Vegetation, Second Edition. Charles D. Bonham.

but often are ignored. The results of data analysis for the single species are biased by the presence of other species and should be corrected.

Three measures of vegetation are available to quantify species characteristics individually or in combination: cover, density, and biomass of plant species in aggregate or individually. Sometimes a fourth measure, frequency, is included, as has been in this book. Frequency is often thought to be a measure of a species characteristic when in fact it is not; it is an indicator variable with respect to a species and takes on a value of 1 or 0 when a species is observed (or not) in a sampling unit. In contrast, on a quadrat-by-quadrat basis with a species characteristic such as biomass, the frequency value of 1 indicates the presence of the species with a nonzero value for biomass. Conversely, a frequency value of 0 indicates absence of the species and subsequently, the biomass value should be zero for that sampling unit.

The use of frequency data to predict biomass on a quadrat–by-quadrat basis results in use of totals or averages for both frequency (x) and biomass (y) obtained from a spatial sampling design that clusters quadrats onto line segments within the area. Cluster sampling or two-stage sampling also can be used to obtain pair-wise estimates of averages for regression/correlation analysis. Replicated line transects with quadrats placed in some fashion will provide multiple estimates for frequency counts or percentages in conjunction with biomass estimates. The point is that the sampling design method must be formulated before choosing the sampling unit (quadrats, lines, etc.)

For practical reasons, without affecting emphasis on a single species, consideration can be given to classifying vegetation in the study area into homogeneous units according to vegetation types, ecological sites, or plant communities. Vegetation types are usually formed on the basis of one or two species that are dominant or codominant, such as grassland, forest, etc. Dominance is usually determined on the basis of species size and abundance. Site characteristics are then often described after vegetation has been characterized. On the other hand, an ecological site is a unit of land with specific physical characteristics and has the potential to support a specific plant association. Several ecological sites may be found within a given vegetation type. Terms such as "range site" and "habitat type" are frequently used in the literature to describe vegetation useful for livestock or wildlife, respectively. "Range site" and "habitat type" are similar, but are not identical in concept. The term "ecological site" embodies the concept of both "range site" and "habitat type" and may be preferred by fieldworkers. All discussion in this chapter applies to vegetation types, plant communities, ecological sites, or other ecologically based classifications.

9.1 Mapping units

There are several ways to prepare a vegetation map showing ecological sites, for example. It is beyond the scope of this book to describe in detail the cartographic techniques for vegetation mapping; however, a brief overview of the technique is described.

9.1.1 Conventional mapping

One should be able to develop a precise vegetation map by using topographic maps, soil survey maps, and aerial photographs. Color photographs provide detailed information on both soil and vegetation and are useful for detailed vegetation mapping. The first step is to delineate general vegetation types on aerial photographs obtained for the study area. Each vegetation type has a characteristic spectral signature that is shown by the tone and texture portrayed by photographs. A reconnaissance field survey is then undertaken to become familiar with different vegetation types. During the survey, a list is made of all plant species occurring within the area and descriptions are made of topography, aspect and slope. Boundaries of vegetation types, as they occur on the ground, are delineated on the topographic survey maps. Soil survey maps also are useful in delineating the boundaries of vegetation types because such boundaries often follow those of soil types.

The second step is to compare overlays of the vegetation boundaries delineated on aerial photographs to those on topographic sheets. Discrepancies, if any, are corrected through an additional field check. The field-produced boundaries are generally more precise with respect to correct identification of vegetation types. On the other hand, the photo-produced overlays are generally more accurate with respect to the overall pattern and distribution of each vegetation type. After the field check, a composite overlay is prepared by putting together the pieces of corrected overlays of the topographic maps.

9.1.2 Satellite imagery mapping

Köhl *et al.* (2006) is an excellent source for an introduction through advanced application of remote sensing and GIS multiresource forest inventory. A brief overview of some techniques with satellite imagery is described here. Use of these advanced techniques to measure vegetation characteristics and to monitor changes over time may become more informative than conducting all field studies via on-ground methodologies. One particular advantage presented through remote sensing and GIS techniques is having a range of the electromagnetic spectrum that correlates with objects on the ground surface; i.e., vegetation components, soil properties, water attributes, and topographic features that can be overlaid on composites with less effort than by conventional techniques.

In this book, Section 4.4.3 provides brief details of how spectral imagery can be used to develop a ground-based field sampling design for data collection and classification of vegetation types within the study area. Satellite imagery from Landsat-5 ETM+ was used by Reich *et al.* (2008) to generate spectral classes for assignment to strata, while additional spectral information and topographic data for the study area were obtained from the Digital Elevation Model (DEM). Digitized maps of soil type are available for many areas. Individual vegetation types or plant communities in many cases have characteristic spectral response bands that can be used in formation of pixel combinations to display as a first draft map of vegetation types and plant communities.

A reconnaissance field survey of the study area is implemented with topographic maps, the satellite imagery, and GPS used to locate field plots of 30×30 m for sampling and monitoring. Without satellite imagery, use a nested plot design, as given in earlier chapters of this book; i.e., nested plots of sizes 10×10 m, 4×4 m, and 1×1 m. This sample design can be developed from the pixel-sized plot divided into nine sub-plots of 10×10 m each. The common precision level for location with a mid-price range GPS is 3 m. There are two ways to improve this precision level: (1) use satellite imagery with a smaller pixel size than the 30 m from Landsat ETM+ and/or (2) use a GPS that provides a more precise location for the coordinates of a pixel. The location of the pixel area is the main focus of using satellite imagery, not the imagery itself for mapping. The purpose is to describe the vegetation, soil, elevation, slope, etc. of each pixel. A plant community of interest may occur within the 900 m^2 or may overlap into adjacent pixel areas of similar spectral characteristics. There is an advantage to using a pixel as a combination of vegetation and its environment at 900 m^2 (<0.25 acre).

There is disagreement among vegetation workers about the use of satellite imagery for field descriptions of vegetation typing and naming of created units. The argument centers around letting pixels from satellite imagery dictate the naming of vegetation units that exist within similar digital pixel boundaries, as opposed to using apparent vegetation types exhibited on aerial photographs and associated soil maps that have been used for decades to map vegetation. In the USA some field ecologists have preferred to continue to conduct surveys on large areas covered by natural vegetation using existing vegetation type maps and find plant communities within these types by field examination.

There are several reasons one should consider the use of satellite imagery to describe an area of ground occupied by natural vegetation. Satellite pixel coordinates are geo-referenced to ground locations and can be used to locate permanent plots for inventory and long-term monitoring. Continued location of permanently marked plots by ground survey methods presents problems in finding the exact original location. If satellite pixel coordinate locations are used to locate plots for inventory, then GPS can be used to locate them for monitoring.

9.1.3 Naming vegetation units

Vegetation units should be given a name that represents the abiotic and/or biotic features. The abiotic part of the name describes readily recognizable, permanent physical features such as a soil type, a topographic feature, or a combination of the two. Some examples are sandy, silty, clay upland, saline, or meadow sites. Sites that have similar soil and topography may exhibit significant differences in their plant communities due to differences in micro climate. In such a situation, a biotic component is added to the name to differentiate among ecological sites occupying similar soil and topography, but differing in species composition. By convention, the biotic name consists of one or two overstory dominant species in forest communities and one or two understory dominant species of shrubs or herbs. Subnames may be used if there are codominants in a community or the community has more than two layers of vegetation. Examples from typical ecological sites in southern Arizona, USA are (1) *Bouteloua chondrosioides/Mimosa dysocarpa* and (2) *Mimosa dysocarpa-Cassia leptadenia/Bouteloua hirsuta.* The dominant species of the predominant life-form is listed first, followed by the dominant species of the next dominant life-form or understory species. The names of species of two different life-forms or layers are separated by a slash (/) and of the two or more codominants, by a dash (-). A common practice is to omit the abiotic component and to name ecological sites after dominant plant species characteristics. This procedure of naming vegetation units is logical since occurrences of vegetation type, plant community, or an ecological site is the result of individual plant species revealing their environmental preferences.

9.1.4 Describing vegetation units

A good description clearly presents features that characterize a vegetation unit. It should not be heavily oriented toward a single use of the resource such as grazing, forestry, or recreation, but should cover all aspects of multiple resource use. Information for a complete vegetation description for all major herbaceous species should include cover, density, and biomass; cover for all other species; and cover and density for shrubs and trees. When possible, common names, although they are regional or even local, should be listed along with scientific names of all plant species present in the vegetation unit (e.g., ecological site) selected for inventory and monitoring.

9.2 Basic considerations

Monitoring and evaluation of vegetation usually require a detection of change in one or more characteristics of plant species over a time interval. No monitoring effort will succeed unless a basic inventory is made according to the long-term needs of the program. All information necessary for evaluation of changes will need to be identified according to species and specific measures needed. Monitoring of vegetation is not difficult if conducted properly. However, a plan for monitoring must be developed before a baseline inventory is made. Then, measures of vegetation characteristics will be comparable throughout the monitoring period. Once characteristics are selected for measurement, a critical analysis should be made of the plan to determine any weaknesses likely to occur in monitoring efforts. During monitoring efforts if changes need to be made in the measures and/or their measurement, a complete data analysis should be conducted on data collected to that point in time. Continuation of data collection intervals remains the same, but statistical procedures will be different for final results than those planned at the beginning of monitoring efforts.

Specific effort should be made to measure plant characteristics that can be monitored to detect change. For example, plant cover should be determined in such a way that a significant ecological change, even of small magnitude, is observed within a meaningful time interval. In general, biological significance cannot be attached to a 1% change in a species cover, regardless of the original cover value. It is easy to understand that an average cover value of 10% initially can change (increase or decrease) by 10%, leaving an average of 9 or 11%, in which case it would be difficult to suggest an ecologically significant change in the species cover value that has occurred even though there might have been a statistically significant change detected. The measurement method and the field sampling design would have to estimate cover in 1% increments. Then cover class intervals of widths greater than 1% are not acceptable as a method of cover measurement. It follows that the level of precision desired should be determined at the outset of the measurement process. Plant density and biomass data of individual species, or collection by life-forms, must meet the same general principle as that set forth for cover estimates.

Additional problems occur in estimating plant cover, total or individually for species, when points are used to record hits. Basically, the same errors occur as those mentioned regarding class intervals. Field ecologists often do not understand that the basic unit determines the precision acquired for an estimate. The sample space associated with the set of possible data values, as defined by Feller (1968), that can be obtained from a population illustrates the problem. The sample space generated from visually estimated cover with intervals of 10% will have a sample space of (0, 10, 20, ..., 100), the only values that can be observed; and there are no fractions of a percentage cover value. Compare this data set with an estimate obtained from a set of 10 points in a frame. The 10 points provide an estimate of cover to the nearest 10%. Repetitions of the 10 point-frames and averaging do not provide closer estimates than 10%, since the basic unit of measure provides intervals of 10% (one hit out of ten = 10% cover) and the sample space is as before for interval estimates. If a 1% precision level is desirable, then 100 points per unit must be used (one hit out of 100 = 1% cover) with a sample space of (0, 1, 2, 3, ..., 100 point contacts) (Bonham and Clark 2005). Other precision levels are also based on number of points used. For example, a 2% level of precision in measurement of cover requires 50 points; a sample space would then be (2, 4, 6, ..., 100) and so on for cover estimates from 25 points or 200 points. This last sample size of 200 points would estimate cover or proportion of any variable to the nearest 0.5%. A change in percentage of cover can only be detected at the level provided by the measurement process.

Problems in monitoring biomass changes either in total or by species may be avoided by deciding initially which species or groups of species will be monitored. Other decisions include whether to separate total biomass of plants into components of leaf, stem, flowers, roots, and so forth. Moreover, consideration should be given as to whether destructive sampling such as harvesting is proper to estimate changes in biomass.

Density measures are not necessarily easier to make for determination of change in vegetation. Definition of an individual grass, forb, or shrub plant can be a problem. Clear descriptions of an ecological plant unit as an individual should be recorded and followed throughout the monitoring period. Otherwise, data may not be comparable to detect meaningful changes, if any, that may occur.

9.3 Sampling and monitoring vegetation

It is common to conduct vegetation inventories and follow up with continued data collection to characterize a plant community at regular intervals. Regularity in data collection is needed in part because information collected at infrequent intervals is of limited use. Also, inventories conducted at fixed intervals, say every five years, may not be very useful in planning that demands information on the population in alternate years. Thus, a series of small samples taken annually may be more informative. When the same community is sampled repeatedly over time, realistic estimates both of costs and variances can be made and techniques that lead to optimum efficiency of sampling can be used, as demonstrated in Chapters 3 and 4.

The primary sampling unit (psu) was discussed in Chapter 4 and related to satellite imagery pixels, as well as ground-based sampling alone. The LandSat 5 satellite will be removed in the near future and replaced by LandSat 8. Other systems are now in place with differing specifications. It is useful to refer to LandSat 5 for illustration purposes. Whether such imagery is used or not in initial monitoring efforts, it is important to have a field design with plots of (e.g. 30×30 m) a size that corresponds to a pixel for a Landsat image and its geographical coordinates, for example. Then a GPS unit can be used to precisely relocate the plot for repeated sampling to compare individual plots both within and among plots to detect changes, if any occurred. If satellite imagery becomes important to the monitoring program, then its incorporation into the current monitoring efforts will provide access to both the time-lapsed past, the present, and future spectral imagery from a Landsat 8 satellite. Such a comparison, for example, will enable detection of vegetation and/or environmental changes that over time has occurred in the study pixels. The satellite signatures can still be related to on-ground data within the plot, even though no imagery was consulted at the time the study location was inventoried. It is worthwhile to consider that while current economic conditions support frequent on-ground monitoring of an area, future economic conditions may not support such frequent repeated sampling of vegetation and its environment.

Several important points need to be made concerning the monitoring of vegetation on a large area and no spectral imagery will be used. Consider, for example, that the monitoring program will be implemented on a large study area (say 1000 ha) and it has been determined that the area's vegetation should be remeasured on a five-year cycle in which approximately a total of 11 100 plots (N the population size for possible sampling) each of 900 m^2 form the population of the large study area. Of course, stratification of the vegetation into vegetation types (forest, grassland, aquatic, etc.) and subsequently further sub-divisions of these types into plant communities will lead to more efficient sampling design possibilities. These communities can be sampled individually or considered as a part of the whole N if pixel size plots are used as primary sampling units divided into nine sub-plots, as described in Chapter 4 and used by Reich *et al.* (2008). If the five years between resampling is apportioned equally with respect to the sample size, then a 10% sample of the area results in about 1100 plots (psu) needed to be measured after five-year intervals. Further assume that some sampling will be completed each year instead of all 1100 plots sampled in the fifth year of a cycle to obtain an estimate of the total biomass or cover, for example, for the entire study area. The number of plots sampled each year would be about 220 plots to obtain estimates for that year. To complete the sampling design by resampling, all sample plots will be randomly assigned to a remeasurement year in such a way as to ensure that every year a representative sample is selected from the entire study area to represent the biomass or cover within the project area. At the end of the first five-year cycle this process will be repeated. This information can be used to assess changes that are occurring on the psu with respect to selected key indicator variables from 1 m^2, 100 m^2, or a Landsat 5 pixel size plot of approximately 30×30 m area (900 m^2). Reich *et al.* (2008) estimated that a sample size for pixel-sized plots placed over vegetation forest types of Jalisco State in Mexico would range from 300 to 700 to obtain data on cover, basal area, canopy volume etc. for forested communities.

9.3.1 Individual plant approach

Gleason's (1926) individualistic concept needs to be understood as to the ecological function of an individual plant species in plant communities, in ecological units or simply occurring within sampling units randomly located in an area. Individual plants of species can be monitored by use of permanently located plots. An individual may be observed in phenological and reproductive stages that occur over a season or years. Of course, all measurements must be non-destructive and employ methods given in previous chapters. Survival data, reproductive rates, and characteristics of cover, density, and biomass can all be measured for change.

Survival is a Bernoulli variable (values of 0 or 1) because a plant either lives or dies. However, as a series of observations over time, it can be modeled as a multinomial function and it is possible to estimate survival rates. The method of maximum likelihood can be used to obtain best estimates of survival rates and these estimates are then used to formulate a series of more complex models to explain variation in data. The models are tested by analysis of likelihood ratios, and the procedure allows testing of climatic, biotic, and environmental factors for their importance as casual factors in the process of change. An example to demonstrate this approach is given by Gardiner and Norton (1983). Note that any data that has a sample space of (0, 1) can be analyzed by the same procedures.

9.3.2 Multispecies approach

No single vegetation characteristic is the best index of the vegetation condition. However, collect only data that is known to detect a significant change in one or more species. Multivariate data analyses through mathematical, statistical, or graphical techniques can be used to interpret the data set.

9.4 Selection of a monitoring procedure

9.4.1 Vegetation measurements

A well-conceived sampling design is a prerequisite to any monitoring program. The choice of a sampling design is dictated, in part, by the objectives, the vegetation type, the vegetation characteristics to be measured, and the availability of financial and technical resources.

Designs most frequently used are completely random sampling, randomized block, and stratified random designs. Vegetation sampling, for both inventory and monitoring, is usually conducted within vegetation types or sub-types that can be analyzed as strata. Sampling locations within a vegetation unit may be selected at random at each measurement time, or repeated measurements may be obtained on the points selected at random the first time. The position of each point is permanently marked on the ground by GPS after driving steel rods to facilitate relocation of the plots or lines. Areas selected for monitoring studies are called "indicator areas" or "key areas." These areas of vegetation should be those that will respond the quickest to a perturbation of any kind or source. Obviously key areas or indicator areas will vary according to a given perturbation.

Frequency sampling may be conducted from plots along permanently located line transects. The plots may be randomly located along the line transect at each point of observation, but usually these are systematically located. Recall that 100 plots give frequency estimates to the nearest 1%.

Density is one of the least used vegetation characteristics for monitoring vegetation trends. This is obviously the result of time needed to count individuals of a species in a bounded area (a quadrat). Density can be estimated by counting the individuals of species on permanently marked two-dimensional plots (squares or rectangles) or by repeated random sampling. Density of trees and shrubs may be estimated by plotless methods. Both the plot and plotless methods of density sampling have been discussed in detail in Chapter 7.

9.4.2 Spectral imagery techniques

Monitoring and evaluation of vegetation characteristics on a time sequence basis may be accomplished by use of imagery available from ground-based photography and electronic spectral imagery or from airborne and satellite imagery. A video system mounted on low-flying aircraft provides multispectral imagery also. This method is readily available, and imagery processing is inexpensive compared to that obtained from satellite platforms. Images from a video camera can be viewed immediately by playback, and resolution is as small as 0.5 m. Digitizers are available for processing film.

The precision needed to detect certain small changes in vegetation may be difficult to obtain for several reasons. If satellite imagery is used, then a unit of measure ranges from 0.1 to 1.4 ha in size, depending on the sensor and band used. This compares to a French satellite system providing a resolution of 0.01 to 0.04 ha in size. Repeat coverage time of satellite systems varies from 16 to 18 days. Additional factors affecting precision of any image obtained are altitude differences, atmospheric distortions, and topographic position of the observed site. Köhl *et al.* (2007) provide details for image processing to remove most sources that influence interpretation of the imagery.

9.4.3 Weight estimate methods

Techniques of estimating biomass have been described in detail in Chapter 8. A rule of thumb often used is that if vegetation is in good condition (stable), then 50% of forage can be utilized without any adverse effect on the health of the plants. Like most "rules of thumb," this one suffers for lack of a quantitative basis. If the vegetation is not in good health, then proportionately more forage should be left ungrazed to enable the vegetation to recover physiologically. On the other hand, some species may be harvested at an 80% level and not be harmed. Once a decision has been made on the percentage vegetation that can be removed by all known

Table 9.1 Approximate number of individuals per animal unit based on ratios of metabolic weights[a] for mature animals. (Reprinted from Heady H. F. 1975, *Rangeland Management* p. 9.)

Species	Approximate weight		$x^{0.75}$	Ratio	No. per
	lb	kg (x)		98/x	animal unit
Cape buffalo	1200	545	112	0.87	0.9
Bison, cow, eland, horse	1000	455	98	1.00	1.0
Elk, zebra	600	272	67	1.46	1.5
Waterbuck, wildebeest	400	182	50	1.96	2.0
Hartebeest, tope	300	136	40	2.45	2.5
Mule deer	150	68	24	4.08	4.0
Sheep, impala	120	55	20	4.98	5.0
Pronghorn antelope, goat	100	45	17	5.76	6.0
Thomson's gazelle	50	23	10	9.80	10.0
Dikdik	12	5	3	32.67	33.0
Black-tailed jackrabbit	5	2.3	2	49.00	49.0

[a]Weight 0.75 kg.

herbivore sources, and if the daily forage requirements of different types of animal are known, then the grazing capacity can be computed for different vegetation subtypes. Grazing capacity is normally expressed in terms of AUM, and a domestic beef cow is taken as a standard. An AUM is the amount of dry forage consumed by a cow and calf in 1 month (30 days). It is estimated that a cow and calf need 10–13 kg (24–31 lb) air-dry forage per day. Therefore, one AUM is equal, approximately, to 400 kg. The animal unit equivalents of other animals are given by the ratio of the metabolic weight of a cow to the metabolic weight of the other animal. The metabolic weight of an animal is three-quarters the power of its body weight in kilograms (Kleiber 1926). Animal unit equivalents of a number of species are given in Table 9.1. Sample grazing capacity calculations by the weight method are shown in Table 9.2. Note that decimal places are not used because g/m^2 weight is seldom obtained to the nearest 0.1 g. Once again, different vegetation types are used by various herbivores in addition to livestock.

9.5 Models versus measurements for monitoring

Measurement of vegetation for the purpose of monitoring requires a high level of precision and repeated measurements. The availability of financial and technical resources may become serious constraints for acceptable levels of statistical error in measurements. A good solution to the problem could be modeling, but the ability of models to adequately follow field data is limited. A combination of model results and field data, however, frequently yields an estimate with satisfactory statistical errors (Jameson 1986ba). The combination may allow more precise interpretation

Table 9.2 Computation of grazing capacity by weight method (Colorado shortgrass site)[a]

List of species	Clipped green wt. (g/m^2) Quadrat no. 1	2	3	4	5	Ave. green wt. (g/m^2) (A)	Dry:green wt. ratio (B)	Ave. dry wt. (g/m^2) (A × B)
Blue grama (*Bouteloua gracilis*)	38	19	42	24	40	33	0.62	20
Western wheatgrass (*Agropyron smithii*)	11	12	2	14	4	9	0.58	10
Buffalograss (*Buchloe dactyloides*)	2	17	13	21	14	17	0.60	10
Sedges (*Carex* spp.)	8	3	6	–	7	5	0.51	3
Three awn (*Aristida* spp.)	2	5	3	1	–	2	0.64	1
Forbs	10	13	9	12	15	12	0.41	5
Total	71	69	75	72	80	78		49

Vegetation sub-type, loamy plains; vegetation condition, good; area, 750 ha; percentage of biomass that can be utilized, 60%.

$$\text{Biomass per hectare (kg)} = \frac{49\,(\text{g/m}^2) \times 10\,000}{1000} = 490$$

$$\text{Grazing capacity (AUM)} = \frac{\text{biomass (kg/ha)} \times \text{area (ha)} \times \text{utilizable proportion}}{400\,\text{kg}}$$

$$= \frac{490 \times 750 \times 0.60}{400}$$

$$= 551 \text{ AUMs}$$

of the results than is possible from measurements alone. For details of the procedure, interested readers with an appropriate statistical background should refer to the work of Jameson.

Monitoring is a costly procedure, but it is possible to reduce cost by comparing the cost of a small sample taken frequently with the cost of a less frequent large sample. The combination of field data with model results can be used to predict the state of the system in the intervening period. Details of these procedures are beyond the scope of this book, and for details, interested readers should refer to Jameson (1985, 1986b).

9.6 Case studies for remote sensing

Aranha *et al.* (2008) used satellite imagery in association with field methods to develop prediction models for estimating biomass from forested and shrub lands; some of which were recently burned-over. The latter areas were of interest from the regeneration of tree populations. The research was conducted in Vale do Alto Tâmega, an area located in Northern Portugal. Digital aerial photographs from 2005 (color on a 0.5 m ground resolution); topographical data on a scale 1/25000, and Landsat TM image from July 2006 were integrated by GPS previously created by Aranha (1998). A total of 400 sampling plots (200 m^2 area/plot) were established throughout the study areas. The NDVI index was calculated from Landsat TM images by

$$\text{NDVI} = (\text{NIR} - \text{Red})/(\text{NIR} + \text{Red}) \tag{9.1}$$

where NDVI is normalized difference vegetation index and NIR is normalized near infrared.

The NDVI and the forest inventory dataset were integrated to develop spatially explicit above-ground biomass (AGB) estimation models for the two forest-shrub types. Stepwise regression analysis procedures were used. Output products needed were: (1) to develop an algorithm for estimating above-ground shrub biomass and (2) to develop an algorithm for estimating above-ground biomass of young maritime pine trees in regeneration areas after fire.

Methods included using IDRISI32, a raster-based GIS; ArcGis 9.1, a vector-based GIS; ArcPad 6.03, a vector-based GIS for PDA-DGPS (personal digital assistant-differential GPS and pathfinder) and GPS data processing software, and a vegetation data set collected in 400 systematic sampling plots.

The dark object subtraction technique was used to calibrate digital images to reduce haze effect before using satellite data to estimate biophysical variables. Geometric correction was made by the nearest-neighbour method to assign satellite imagery to their correct position on the Earth's surface and coregister them to the Datum Lisboa Hayford-Gauss coordinate system. This system does not average the original values when performing the resample process. Principal components and NDVI values were used create new images with stronger contrast among the pixels. These new images were used, as extra bands, to improve results of image classification. Twelve land cover classes (CORINE land cover classes) were used to classify land cover and to analyze accuracy of the methods. Field data was imported into IDRISI to pair plot biomass data with NDVI plot values.

Predictive biomass models (tons/ha) for shrub land and for *Pinus pinaster* self regeneration areas are

$$\text{Shrub (tons/ha)} = 9.17 + 3.000\ \ln(\text{NDVI}) \tag{9.2}$$

The first term on the right-hand side of the equation is a constant and the numerical value, 3.0 is the change in shrub biomass per unit of change in the value of NDVI. ln is the natural log and adjusted $R^2 = 0.355$ (p-value < 0.01) for the linear fit. The equation for the pine regeneration area is

$$\text{Reg\_Pp (tons/ha)} = 5.96 + 2.557\ \ln(\text{NDVI}) \tag{9.3}$$

Definitions of numerical values on the right-hand side of the equation are the same as for the previous equation. Reg Pp is for "regeneration of the *Pinus pinaster*" area. Adjusted $R^2 = 0.244$ (p-value < 0.05). The authors noted that NDVI values ranged from 0.01 to 0.7.

Poulin *et al.* (2002) concluded the potential and reliability of satellite imagery are still relatively unknown for a large number of ecosystems and this is apparently true, even though imagery may be becoming a basic component of the work done by plant ecologists. These authors conducted a study to determine the reliability of using Landsat 7/ETM+ (Enhanced Thematic Mapper Plus) data to map habitats of peatlands. The image contained radiometric information in six spectrally defined channels in 30×30 m pixels in addition to a thermal infrared radiation channel with 60×60 m pixels and a panchromatic channel of 15×15 m pixels.

Before attempting to classify these habitats, a mask procedure was applied that revealed 626 peatland locations and each location was recorded with a GPS. A habitat class was associated with each of these locations in the field by visual identification. Both a simple maximum likelihood (ML) function and a weighted maximum likelihood (WML) function were used to consider the proportion of each habitat class within each peatland. Both classification procedures provided accurate representations of 13 peatland habitat classes. The ML was better for mapping rare habitats, while WML was better at mapping the most common habitats. The WML procedure was more reliable to identify potential conservation units than was the ML procedure because it provided a more robust identification of rare habitats. In which case, the chance of misallocating protected sites would be minimized and the peatland classification procedures would be beneficial for land managers and conservationists.

Vegetation structure was evaluated in a 20×20 m quadrat for each pixel and was based on 17 strata: water, litter, upright mosses, horizontal mosses, liverworts, lichens, three *Sphagnum* sections (*Acutifolia*, *Cuspidata*, *Sphagnum*), ericaceous shrubs, *Carex*, other sedges and forbs, shrubs, birch, pine, spruce, and larch. The percentage cover of these vegetation strata were visually estimated walking across a 400 m^2 quadrat. Percentage cover was assigned to one of seven classes: present, 1.5, 6.10, 11.25, 26.50, 51.75, 76, and 100% with mid-points being used for statistical analyses. Plant species composition was estimated to the nearest 1% in three circular plots of 0.65 m^2 non-systematically distributed within a sampled pixel.

Canonical correspondence analysis (CCA) and redundancy analysis (RDA) were used to model the relationship between species composition and explanatory variables, such as habitat classes, and environmental and spatial variables, respectively. Ordinations were computed using Canoco 4.0 (ter Braak and Milauer 1998). The ordination was conducted from the results of a detrended correspondence analysis (DCA) that showed a large gradient for species data. Other topics discussed in detail included validation constraints and vegetation patterns.

In general, habitat classes reflected ecological attributes that are relevant to the land management and conservation problems that exist. The ordination showed that peatland habitats that were defined *a priori* for the supervised classification procedures were representative of the species distribution and vegetation structure patterns in the peatlands of the mid-St. Lawrence plain of southern Quebec, Canada. The authors felt that the most important outcome of their research was that their results are readily accessible to land planners and conservationists seeking to establish nature reserve networks.

9.7 Plant species evaluations

Plant ecologists of the present era still seek to develop a deeper understanding of the ecology of plants in their environment. Statistical approaches do not suffice in many cases, while other quantitative methods from the early days of plant ecological studies might do so. Reference works published in the first decade of 2000 in scientific journals and in books still present statistical approaches to the analysis of field data. An exception is a recent reference that offers an in-depth approach, in a summary and explanations of many methods used by plant ecologists over past decades and still used by present-day ecologists. Materials presented in this measurements book have emphasized measurements, not data analysis, and no attempt was made to supplement materials that do provide detailed analysis methods. The following materials provide exclusively quantitative methods for plant ecologists.

Wildi (2010) presents a compendium of methods to analyze plant and vegetation (as a group of species) data. Topics include patterns in vegetation ecology, principal component analysis (PCA), ordination (graphical representation of sampling units, species, stands for sampling major species), and ordering or ordination of these groups using similarity coefficients (species and community similarity coefficients). In addition to methods for analysis of species and groups of species and their environmental factors, Wildi (2010) presents detailed descriptions of classification methods for species, stands, and so forth to study how species behave, for example, in a continuous, not really discrete manner in groupings of plants. Other topics covered by Wildi (2010) make for an up-to-date (as of 2010) account of methods for quantitative plant ecologists.

9.7.1 Plant association

This important work covers the major methods used by plant ecologists seeking to interpret their field data in terms of vegetation–environment relationships. A contingency test is used for the number of joint occurrence of pair-wise plant species in

Table 9.3 Data summaries for Contingency analysis for pair-wise species/communities

Plant Species of Community B	Plant Species of Community A		
	+	−	
	a	b	* a + b
	c	d	* c + d
	a + c	b + d	* Total

two plant communities. Gleason (1925) was apparently the first to suggest the use of a contingency table to determine the probability of joint occurrence between two species, but he did not make a specific test for significance of departure from the expected values. Wildi (2010) listed tests such as Jaccard's coefficient, Sorensen's coefficient, or Fisher's contingency table (count data) can be used. The analysis has the same basic structure with data placed in a 2 × 2 table of species occurrences and their joint occurrences in two plant communities or ecological sites (Table 9.3), where a is the number of species that occur in both communities (A and B). The number of times a species occurs in one community only is b or c while d is the number of species that did not occur in either community. The row sums of $a + b$ is the number of species occurring in community B, $a + c$ is the sum of species occurring in community A and the total of occurrences for both species is the sum $(a + b + c + d)$, T. Most plant ecologists have experienced such a table in basic statistics where the χ^2 (chi-squared) test is used to statistically analyze count data for plant species occurrences in pair-wise communities or in quadrats within a plant community. A still popular index considered to be a distance measure of maximum percentage of association (species affinity index, SAI) of any two species over the entire study is estimated by Czekanowski's index (Bray and Curtis 1957)

$$\text{SAI} = \frac{2\,w}{a + b} \tag{9.4}$$

where $w = \sum_{i=1}^{n} m_i$; n = number of stands sampled, m_i = the lesser occurrence of either species A or B in the ith stand, a is the number of occurrences of species a, and b is the total occurrences of species b in the study. An ordination of species is then calculated using these values. The ordination points on the graph for each species is computed by the formula $(X) - (Y)$, where (X) = the coefficient for each species calculated with species X, which was selected as an end-point, and (Y) = the coefficient for each species calculated with species Y, the other end-point. A positive value (+) sign indicates closer affinity of a species to (X) and a negative sign (–), indicates a closer affinity to species (Y), using zero (0) as a reference point.

An ordination of stands can be conducted for comparison of pair-wise stands. Equation (9.4) is used, but the variables differ in definition and follow that of Curtis (1959) accordingly, $w = \sum_{i=1}^{n} m_i$; n = number of species encountered, m = the lesser occurrence of the ith species in stand A or stand B, A = total occurrences of all species in stand A, and B = total occurrences of all species in stand B. Calculations and points to be represented on a graph follow the same procedure as for the species above.

Dice (1945) used presence/absence data from prairie vegetation to determine interspecific correlation for species. He suggested the "association index" (AI) to measure association between two species. This index is also used in ordination of pair-wise species to interpret their responses to each other and their environment. To compare species A to species B, let the index I have a value for A to B equal to

$$\mathrm{AI} = \frac{h}{b} \tag{9.5}$$

where h = number of joint occurrences in quadrats, a = number of times species A occurred, and b = number of times B occurred

$$\mathrm{BI} = \frac{h}{a}.$$

9.7.2 Diversity indices

There are several measures of diversity of species within plant communities, stands, ecological types, and so forth. The meaning of diversity varies in the literature, but essentially the two most common are: (1) the total number of species in a community, but is best described as species richness, and (2) the dual concept of diversity, which combines species richness and the relative abundances of species (Peet 1974 and 1975). These references address the topics of measurement of diversity and the relative diversity indices that can be used in studies of plant associations.

There are a number of equations used to place numerical values on a measure of species diversity. Of the many, The Shannon–Weaver index is the mostly widely used (Ludwig and Reynolds 1988). The Shannon function is calculated as

$$H' = -\sum_{i=1}^{S}\left[\left(\frac{n_i}{n}\right)\ln\left(\frac{n_i}{n}\right)\right] \tag{9.6}$$

where H' is a sample estimate of the uncertainty per species in the community, n_i the number of individuals belonging to the ith species (S) in the sample and n is the total number of individuals in the sample. This equation is the most used form of the Shannon index in the ecological literature (Ludwig and Reynolds 1988). Like so many estimates obtained to estimate parameters of equations used in ecology, this one also is biased unless n is large (i.e., $n \geq 30$). If cover is estimated to the nearest

1% then 100 quadrats or points are needed, or to the nearest 2%, 50 sampling units are needed. Then $n = 100$ or 50, depending on the precision needed for an estimate of H'.

9.8 Bibliography

Aranha J. (1998). An Integrated Geographical Information System for the Vale do Alto Tâmega. PhD Thesis, School of Geography, Kingston University: London.

Aranha J. T., Viana H. F., and Rodrigues R. 2008. Vegetation classification and quantification by satellite image processing. A case study in North Portugal. In *Bioenergy: Challenges and Opportunities, International Conference and Exhibition on Bioenergy, April 2008*. Univerersity do Minho: Guimarães, Portugal.

Bonham C. D. 1987. Estimation of forage removal by rangeland pests. In Caperina J. L. (ed.), *Integrated Pest Management on Rangeland – A Shortgrass Prairie Perspective*. Westview Press: Boulder, CO.

Bonham C. D. and Clark D. L. 2005. Quantification of plant cover estimates. *Grassland Science* **51**: 129–137.

Bray J. R. and Curtis J. T. 1957. An ordination of the upland forest communities of southern Wisconsin. *Ecol. Monogr.* **27**: 325–349.

Curtis J. T. 1959. *The Vegetation of Wisconsin: An Ordination of Plant Communities*. University of Wisconsin Press: Madison, WI.

Dice L. R. 1945. Measures of the amount of ecological association between species. *Ecology* **26**: 297–302.

Feller W. 1968. *An Introduction to Probability Theory and its Applications.* John Wiley & Sons, Inc.: New York.

Gardiner H. G. and Norton B. E. 1983. Do traditional methods provide a reliable measure of range trend. In Bell J. F. and Atterbury T. (eds.) *Proceedings of International Conference on Renewable Resource Inventories for Monitoring Changes and Trends (August 15–19, 1983)*, Oregon State University: Corvallis, OR, 618–622.

Gleason H. A. 1925. Species and area. *Ecology* **6**: 66–74.

Gleason H. A. 1926. The individualistic concept of the plant association. *Bull. Torrey Bot. Club* **53**: 7–26.

Heady, H. F. 1975, *Rangeland Management*. McGraw-Hill: New York.

Kleiber M. 1961. *The Fire of Life: An Introduction to Animal Energetics*. John Wiley & Sons, Inc.: New York.

Köhl M., Magnussen S., and Marchetti M. 2006. *Sampling Methods, Remote Sensing and GIS Multiresource Forest Inventory*. Springer-Verlag: Berlin, Heidelberg, New York.

Jameson D. A. 1985. A priori analysis of allowable interval between measurements as a test of model validity. *Appl. Math. Comput.* **17**: 93–105.

Jameson D. A. 1986a. Models versus measurements in grazing systems analysis, pp. 37-41. In *Symposium Proceedings. Statistical Analysis and Modelling of Grazing Systems Data* (February 9–16, 1986), C. D. Bonham, S. S. Coleman, C. E. Lewis, and G. W. Tanner (eds). Society of Range Management 39th Annual Meeting, Orlando, FL.

Jameson D. A. 1986b. Sampling intensity for monitoring of environmental systems. *Appl. Math. Comput.* **18**: 71–76.

Ludwig J. A. and Reynolds J. R. 1988. *Statistical Ecology*. John Wiley & Sons, Inc: New York.

Peet R. K. 1974. The measurement of species diversity. *Ann. Rev. Ecol. Syst* **5**: 285–307.

Peet R. K. 1975. Relative diversity indices. *Ecology* **56**: 496–498.

Poulin M., Careau D., Rochefort L., and Desrochers A. 2002. From satellite imagery to peatland vegetation diversity: how reliable are habitat maps? *Conserv. Ecol.* **6**: 16.

Reich R. M., Aguirre-Bravo C., and Briseno M. A. M. 2008. An innovative approach to inventory and monitoring of natural resources in the Mexican State of Jalisco. *Environ. Monitor. Assess.* **146**: 383–396.

Shimwell D. W. 1971. *The Description and Classification of Vegetation*. University of Washington: Seattle.

Ter Braak C. J. F. and Smilauer P. 1998. *Canoco Reference Manual and User's Guide to Canoco for Windows: Software for Canonical Community Ordination* (version 4.0). Microcomputer Power: Ithaca, NY.

Whittaker R. H. 1951. A criticism of the plant association and climatic climax concepts. *Northwest Sci.* **25**: 17–31.

Whittaker R. H. 1967. Gradient analysis of vegetation. *Biol. Rev.* **49**: 207–264.

Wildi O. 2010. *Data Analysis in Vegetation Ecology*. John Wiley & Sons, Ltd: Chichester, UK.

Appendix: Unit conversion tables

Table A.1 Conversion constants from the English system to metric units

From			To	
		Length		
in	inches	2.54	centimeters	cm
ft	feet	0.30	meters	m
mi	miles	1.6	kilometers	km
		Area		
in^2	square inches	6.5	square centimeters	cm^2
ft^2	square feet	0.09	square meters	m^2
mi^2	square miles	2.6	square kilometers	km^2
ac	acres	0.4	hectares	ha
		Mass (Weight)		
oz	ounces	28	grams	g
lb	pounds	0.45	kilograms	kg
		Volume		
fl oz	fluid ounces	30	milliliters	ml
pt	pints	0.47	liters	l
qt	quarts	0.95	liters	l
gal	gallons	3.8	liters	l
ft^3	cubic feet	0.03	cubic meters	m^3

Measurements for Terrestrial Vegetation, Second Edition. Charles D. Bonham.

Table A.2 Constants for converting weight/small area to weight/large area

From g/area	Constant	To
0.1 m^2	10	g/m^2
0.5 m^2	2	g/m^2
0.1 m^2	100	kg/ha
0.25 m^2	40	kg/ha
0.5 m^2	20	kg/ha
1.0 m^2	10	kg/ha
0.1 m^2	89.2	lb/acre
0.25 m^2	35.7	lb/acre
0.50 m^2	17.8	lb/acre
1 m^2	8.9	lb/acre
0.96 ft^2	100	lb/acre
9.6 ft^2	10	lb/acre
96 ft^2	1	lb/acre

Index

Measurements for Terrestrial Vegetation, Second Edition. Charles D. Bonham.

Printed and bound by CPI Group (UK) Ltd, Croydon, CR0 4YY

09/10/2024

14571435-0002